SAE EDGE™
RESEARCH REPORT

Unsettled Issues Regarding Policy Aspects of Automated Driving Systems

Bob McQueen
Bob McQueen and Associates

EDGE DEVELOPMENT TEAM

Vik Bhide, *City of Tampa, Florida*
Doug Jamison, *Central Florida Regional Transportation Authority*
Jeremy Dilmore, *Florida Department of Transportation*
Peter Thompson, *San Diego Association of Governments*
Barrie Kirk, *Canadian Automated Vehicles Centre of Excellence (CAVCOE)*
TJ Ross, *Pace Suburban Bus (Ret.)*
James Garner, *Pace Suburban Bus*
Peeter Kivestu, *Oplytix*

SAE INTERNATIONAL®
Warrendale, Pennsylvania, USA

INTERNATIONAL®

About the Publisher

SAE International® is a global association of more than 128,000 engineers and related technical experts in the aerospace, automotive, and commercial vehicle industries. Our core competencies are life-long learning and voluntary consensus standards development. Visit sae.org

SAE EDGE™ Research Report Disclaimer

SAE EDGE™ Research Reports focus on topics that are dynamic, in which knowledge is incomplete, and which have yet to be standardized. They represent the collective wisdom of a group of experts and serve as a practical guide to the reader in understanding unsettled subject matter. They are not meant to provide a recommended practice or protocol. The experts have assembled as a community of practitioners to contribute and collectivize their thoughts and points of view; these are not the positions of the institutions or businesses with which they are affiliated, nor is one contributor's perspective advanced over other contributors. SAE EDGE™ Research Reports are the property of SAE International and SAE alone is responsible for their content.

About This Publication

SAE EDGE™ Research Reports provide state-of-the-art and state-of-industry examinations of the most significant topics in mobility engineering. SAE EDGE™ contributors are experts from research, academia, and industry who have come together to explore and define the most critical advancements, challenges, and future direction in areas such as vehicle automation, unmanned aircraft, cybersecurity, advanced propulsion, advanced manufacturing, Internet of Things, and connectivity.

Related Resources

SAE MOBILUS® Automated & Connected Knowledge Hub
https://saemobilus.sae.org/automated-connected

SAE Team

Frank Menchaca, Chief Growth Officer
Michael Thompson, Director of Standards, Information and Research Publications
Monica Nogueira, Director of Content Acquisition
Beth Ellen Dibeler, Product Manager
William Kucinski, Managing Technical Editor

Printed in USA

EPR2020016
ISSN 2640-3536
e-ISSN 2640-3544
ISBN 978-1-4686-0204-3

To purchase bulk quantities, please contact: SAE Customer Service

E-mail: CustomerService@sae.org
Phone: 877-606-7323 (*inside USA and Canada*)
+1-724-776-4970 (*outside USA*)
Fax: +1-724-776-0790

https://www.sae.org/publications/edge-research-reports

About the Editor

Bob McQueen is an internationally recognized expert in the application of advanced technology to transportation. He has provided expert-level consulting advice to central and local governments in the Middle East, Asia-Pacific, Europe, and North America. He has also advised global companies when entering or improving positioning in the market for advanced transportation solutions. McQueen specializes in the concise communication of complex technology concepts, matching user needs to technology capabilities and constraints. McQueen has had the privilege of gaining in-depth experience in technology planning, interactive requirements analysis, international standardization, and the application of big data and analytics. He is Chief Executive Officer for Bob McQueen and Associates, based in Orlando, Florida.

contents

Unsettled Issues Regarding Policy Aspects of Automated Driving Systems

Abstract

Automated driving systems (ADS) represent an area of considerable investment and activity within the transportation sphere. The potential impact of ADS on safety, efficiency, and user experience are extremely significant. To get the most from the technology, it is important to ensure that policies are developed to support the balance between achieving public sector objectives and supporting private sector innovation. This SAE EDGE™ Research Report explores the policy aspects related to ADS technology, explains the key stakeholders, identifies unsettled issues, and proposes a number of steps to move forward and improve the current situation. It is hoped that the report will provide a valuable resource to those involved in the definition of ADS policy from both public and private perspectives. It is also intended to serve as a resource for those involved in ADS planning and development and public sector staff involved in other aspects beyond ADS policy.

NOTE: SAE EDGE™ Research Reports are intended to identify and illuminate key issues in emerging, but still unsettled, technologies of interest to the mobility industry. The goal of SAE EDGE™ Research Reports is to stimulate discussion and work in the hope of promoting and speeding resolution of identified issues. SAE EDGE™ Research Reports are not intended to resolve the challenges they identify or close any topic to further scrutiny.

BOB MCQUEEN

Bob McQueen and Associates

Edge Development Team

Vik Bhide, *City of Tampa, Florida*
Doug Jamison, *Central Florida Regional Transportation Authority*
Jeremy Dilmore, *Florida Department of Transportation*
Peter Thompson, *San Diego Association of Governments*
Barrie Kirk, *Canadian Automated Vehicles Centre of Excellence (CAVCOE)*
TJ Ross, *Pace Suburban Bus (Ret.)*
James Garner, *Pace Suburban Bus*
Peeter Kivestu, *Oplytix*

ISSN 2640-3536

Introduction

State of the Industry

In 1865, the British government introduced legislation known as the Locomotives on Highways Act of 1865 [1]. This was one of a series of three pieces of legislation relating to road vehicles. The 1865 Act required all road locomotives, including automobiles, to travel at a maximum of 4 mph in the country and 2 mph in the city. It also specified that a man carrying a red flag was required to walk in front of the vehicle while it was in operation. While the legislation imposed severe constraints on the operation of road vehicles, it also enabled the automotive industry in the UK to develop soon after the introduction of the first practical production automobile by Karl Benz, in Germany, in 1885. Learning lessons from history, it is apparent that policy and legislation can have a significant impact on progress to introduce new technology to transportation. Policy can act as a constraint and an enabler. Another lesson from 1865 is that in the face of new technology and change, government reaction can tend to be extreme, especially with respect to public safety.

Significant progress has been made in the development of automated driving system (ADS) technology. In fact, it can be argued that the era of the ADS is already here. Thousands of pilot projects and demonstrations are underway, although the market is not quite established for on-road use of ADS-equipped vehicles, also known as automated vehicles (AVs). It is interesting to note that market establishment is much further advanced for off-road applications in mining and agriculture. A combination of sensors, hardware, and software can replace the human driver and operate vehicles in a fully automated manner. While a gradual, step-by-step approach to the automation of the driving task is being adopted, full automation is now on the horizon.

Policy Aspects of ADS Technology

The intent of this SAE EDGE Research Report is to provide a resource that will support policymakers in developing effective and balanced policies that take account of the need to protect public safety and the public interest, while also encouraging the establishment and growth of an ADS and AV industry. The intended audience for the report includes those involved in both the public and private sectors in planning, design, and implementation activities related to AVs. It is worth noting that "unsettled" does not infer that there is no existing body of knowledge or extensive current activity with respect to ADS policy. What it does convey is that additional work is still required in order to add to the body of knowledge, address all relevant issues, and take full advantage of progress to date. In recognition of the substantial body of knowledge that already exists with respect to ADS policy, a section has been included in the report that provides sources of further information from the current body of knowledge.

What is an ADS?

While ADS characteristics and taxonomy are explained later, it is useful to provide an overview to prepare the ground for the rest of the report. ADS is a technology application consisting of components and systems that replaces the driving function and enables vehicles to be driven automatically with no human intervention.

SAE International has defined five levels of automation to create a roadmap from the current manually driven vehicle situation to full automation in the future (Figure 1).

Level 5 in the diagram is the focus for the research work. Policy aspects related to full automation of the driving function are explored. Private cars, transit vehicles, and freight vehicles are also included in the research, including various forms of transit including fixed route, on-demand, and first mile/last.

What Is Policy?

While policymakers will already possess a detailed understanding of the definition of policy, to support an effective discussion on policy with non-policymakers it is helpful to provide a working definition for the term. The definition used in this report-taken from the Merriam-Webster dictionary [2] states that policy can be defined as follows:

- Prudence or wisdom in the management of affairs
- Management or procedure based primarily on material interest
- A definite course or method of action selected from among alternatives and in light of given conditions to guide and determine present and future decisions
- A high-level overall plan embracing the general goals and acceptable procedures especially of a governmental body

Considering each of these in turn from both public and private perspectives, further explanation is provided as follows:

- **Prudence or wisdom in the management of affairs**

From a public sector perspective, this indicates the need for a balance between risk management and support for innovation in the application of technology. Most public sector policy objectives require a compromise between multiple goals. Wisdom in the management of affairs also involves the effective engagement of the private sector to ensure good stewardship of public funding, avoiding duplication of effort, and taking full advantage of private sector motivation and

FIGURE 1. SAE levels of automation.

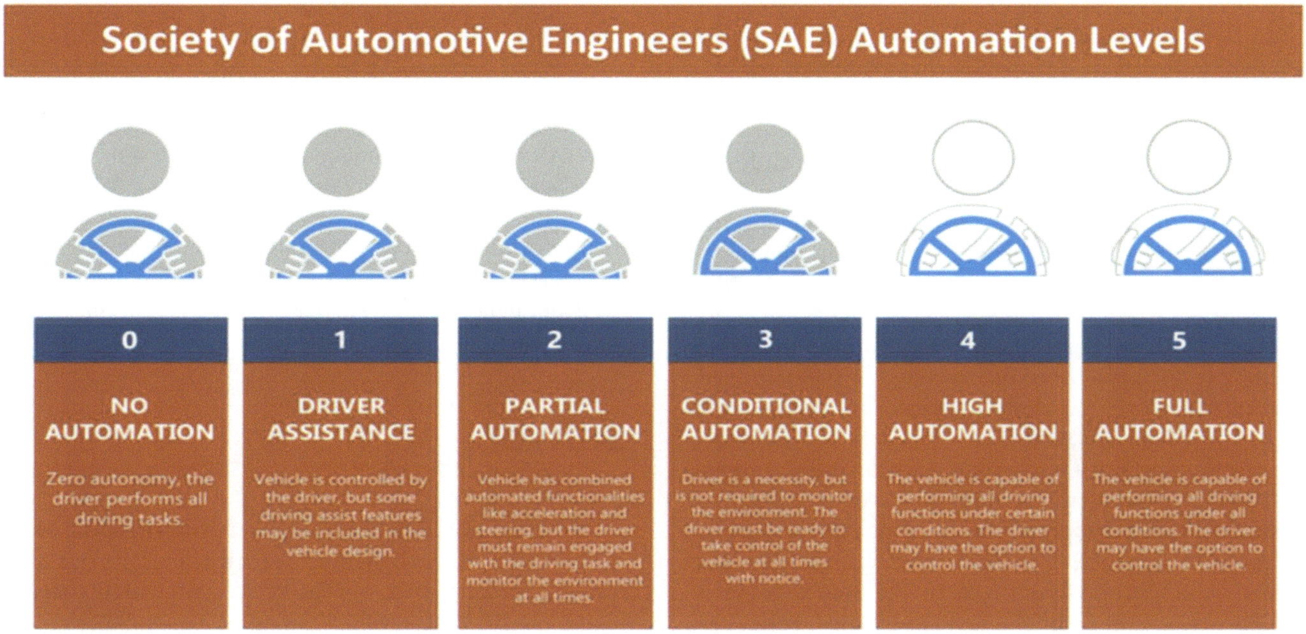

investments. This is particularly important as the current ADS/AV trend is for the private sector to lead through investments (to create markets, products, and services).

From a private sector perspective, prudence or wisdom in the management of affairs involves the protection of sunk investment in the development of new products and services. It also includes wise investments based on knowledge of the market and customer needs, as well as public initiatives, policies, and regulation.

- **Management or procedure based primarily on material interest**

The primary interest for the public sector is attaining the full benefits from ADS application in the transportation network. Procedures, or policies, are designed to guide public effort and provide information and guidance to the private sector. Policy can also be accompanied by regulation, which provides a more formal set of instructions and can form the basis for laws that require compliance.

From a private sector perspective, primary interest can be defined as the ability to create a market for products and services, develop sustainable business, and provide a reasonable return for investors and stakeholders.

- **A definite course or method of action selected from among alternatives, and in light of given conditions, to guide and determine present and future decisio**ns

This also refers to the need to achieve a balance from among alternatives, taking account of current and future conditions. From the public sector perspective, policies align with political and transportation objectives and describe a course of action designed to achieve objectives in the most effective manner. It does so while balancing the protection of public interest, with the desire to enable innovation and rapid product and service introduction. Given conditions can also include the state of current and emerging technology. In some cases, the availability of innovative technology requires a policy response—in other cases, the definition of a policy to shape and guide the emergence of technology.

From the private sector perspective, this course of action would take the form of a business plan that describes strategies and tactics to be utilized to address the ADS market.

- **A high-level overall plan embracing the general goals and acceptable procedures especially of a governmental body**

From a public sector perspective, this involves the definition of a high-level overall plan that includes the alignment of policies with published transportation plans. Policies must be aligned to agreed plans and proposed investments, to achieve the maximum effect. Policies can be used to guide the plan, or if the plan has already been published, the plan will guide policy. This part of the definition also includes important information on not just the public sector goals, but the acceptable procedures that can be applied to achieve the goals.

From a private sector perspective, it will be necessary to understand public sector plans and acceptable procedures, while implementing enterprise-specific plans and processes. Alignment between the public and private sector is vital if the best results are to be achieved.

Why Policy Is Important

The importance of policy lies in the results or effects on the public and private sector enterprises. There are two extremes that can describe policy effects. At one extreme, policy can stifle innovation through the rigid definition of a set of rules that focus on risk management, while not allowing innovation. At the other extreme, policy can have the effect of stimulating ADS investment and encouraging the rapid introduction of new and emerging technologies to the detriment of safety. It is crucial to consider both aspects.

Policy is also an important communication tool from the public to the private sector. As stated in the definition of policy, it involves the development of procedures and guidance based on the attainment of public sector objectives while minimizing risk. Good policy will take full account of technology capabilities, the needs of the traveling public, and provide an optimized approach to achieving what is required and desired-while avoiding undesirable effects. Policy has the potential to support a vibrant ADS market that will enable the full benefits of the technologies to be released. Policy can protect both public and private sector interests by ensuring that public objectives are clearly stated and communicated to the private sector.

One of the crucial aspects of policy related to ADS technology will be the close alignment of policies from various levels of government to those of the private sector.

ADS Terminology and Taxonomy

To enable an effective discussion of ADS policy, it is important to have a clear understanding of the nature and characteristics of ADS technology, terminology, and taxonomy.

There are significant variations in specific ADS designs, including the use of different sensor types and overall architectures for linking components together. However, a high-level system architecture can be defined that characterizes the general nature of an ADS and its individual components (Figure 2).

Sensors

LIDAR A light distancing and ranging (LIDAR) sensor emits beams of laser light and measures how long it takes for the light to return to the sensor. This enables a highly accurate digital picture of the vehicle surroundings to be developed. It includes obstacles, people, and the topography of the road network.

RADAR A radio distance and ranging—or radar—sensor emits radio waves and measures the time taken for the radio wave, reflected from objects in the environment, to return to

FIGURE 2. High-level architecture Automated Driving Systems [3].

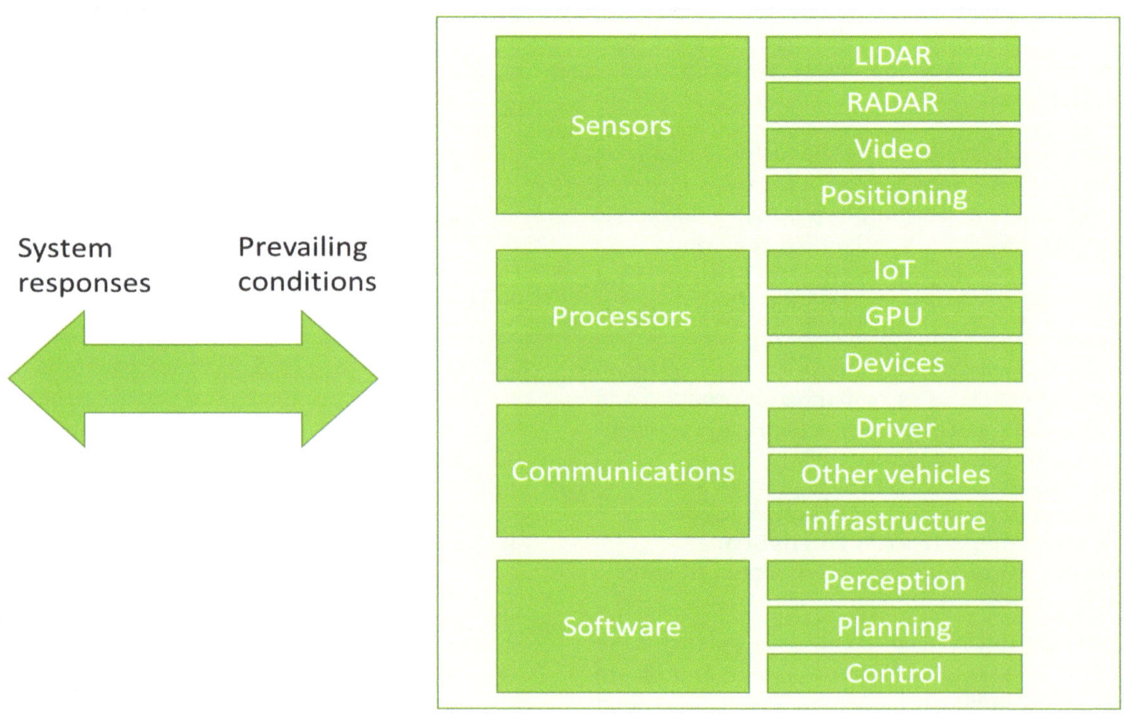

the sensor. In addition to recent ADS applications, radar has been in use since World War II for object detection, vehicle speed measurement, weather measurement, missile and aircraft guidance, and air traffic control.

Video High-resolution video cameras complement data input from LIDAR and radar sensors. Video cameras provide images, which are then processed and converted into digital data.

Positioning Positioning sensors use a range of technologies including satellite location and inertial navigation techniques to establish the precise location of the vehicle, direction of travel, vehicle speed, and other parameters that describe position and geometry.

Processors

Internet of Things Internet of Things (IoT) processors manage data processing and connectivity for the outside world. The name originates from the expansion of the Internet to connect devices and processors as well as people and computers.

Graphics Processing Units Specialized graphics processing units manipulate image data and support high-speed mathematical calculations for image processing.

Devices Other onboard vehicle devices used to address ADS processing functions may monitor temperature, tire inflation, oil pressure and temperature, and other vehicle parameters.

Communications

Driver These devices include voice-activated and visual techniques enable communication with the ADS. In some cases, tactile techniques have also been used to communicate to the driver.

Other Vehicles Wireless communication techniques are used to support connectivity between the AV and other suitably equipped vehicles. Telecommunication technology utilized can include cellular wireless, dedicated short-range communications, and specialized microwave communications.

Infrastructure Communication between the vehicle and infrastructure is also supported by the same communication technologies used to communicate with other vehicles. These can include cellular wireless, dedicated short-range communications, and specialized microwave communications.

Software

Perception Perception software analyzes sensor data to understand the environment surrounding the vehicle, vehicle performance, and driver behavior. Receiving inputs from the sensors onboard the vehicle, the software supports perception functions, such as localization in which the three-dimensional (3D) location of the vehicle is established in relation to a digital map. This also includes 3D orientation.

Planning Planning software involves the definition of proposed actions in response to data inputs, including decision-making algorithms to recognize scenarios and make appropriate decisions. It will also include regression algorithms used for object detection, localization, and prediction. Pattern-recognition algorithms will also be used for object detection and recognition.

Control Control software activates vehicle systems and actuators such as steering, brakes, and accelerator.

Note that all the above components are integrated into a single ADS application in which they work together in a coordinated and harmonious fashion. As indicated earlier, there are considerable design variations between different products and services, and high-level architecture illustrated represents a generic representation designed to explain the aspects of ADS technology.

Aspects of Policy

While this section will not include an exhaustive list of ADS policy aspects, it will help the target audience by providing a centralized resource describing policy aspects based on the current state of policymaking. ADS technology represents a multidimensional challenge with respect to policy. In most cases, policy must be customized to national and local needs. However, the policymaking process can be made as efficient as possible by referring to existing policies and lessons learned to date. This section of the report supports this by providing a checklist to be used as a starting point for policy formulation.

This also illuminates the importance of policy by shedding light on the different aspects and the impacts policy will have on transportation and the general population (Table 1). While the public sector is expected to lead the development of policy, there will be effects and impacts on the private sector related to detail in the policy. This will be explored in later sections.

Data

Data elements of an effective policy will address data sharing, data privacy, and data protection. It is important to define what will be done, and what will not be done with the data,

TABLE 1. Policy aspects.

No.	Aspect
1	Data
2	Driver
3	Economy
4	Efficiency
5	Energy
6	Environment
7	Ethics
8	Infrastructure
9	Insurance
10	Land-use and urban planning
11	Liability
12	Mobility, connectivity, accessibility
13	Modes
14	Operational
15	Partnership
16	Procurement
17	Regulation
18	Public health
19	Safety
20	Security
21	Social equity
22	Standards
23	Technology
24	Telecommunications
25	Testing
26	Vehicle
27	Workforce development

including the definition of suitable arrangements for data governance. Relating to data sharing, the policy should articulate proposals for sharing data and arrangements for secure transfer. Relating to data governance, the arrangements to control access to the data, maintain the appropriate level of privacy, and preserve security should also be defined. The ownership of data is also a significant part of this policy aspect. It may require a more detailed approach to the management of data sourcing, data combination, and the creation of information. This could lend itself to a data monetization approach that considers both the public and private sectors' value of data and the information derived from the data. This is an important consideration due to the expense of collecting and managing data, and the opportunity to understand the value of the information that is derived from the data.

Driver

This may seem a little strange as the driver is replaced by an ADS. However, in the short term, it is expected that there will be a transition from no-automation through partial automation—or driver decision support-to full automation.

Policy aspects related to the driver will include compliance with prevailing traffic laws and regulations and the interaction between the driver and the ADS. With respect to transit, the policy will also address passenger interactions with ADS technology as well as bus operators. This policy aspect will have a significant impact on the driver with respect to the enablement of ubiquitous AVs. It is expected that this will have a significant effect in improving safety by reducing accidents and fatalities. It will also affect the driver through the availability of other transportation options including transit services and automated "first mile, last mile" shuttles. Policy will dictate the availability of such services and the ability to operate them on public roads and highways.

For transit applications, the role of the driver or operator goes beyond the tasks involved in driving the vehicle to include tasks related to safety, security, local knowledge and traveler information, spills and incident reporting, fare collection compliance, response to incidents, and personal interaction with travelers. The driver also acts as a supervisor for the vehicle. Currently, many parents rely on this to safely and securely send unaccompanied young children by public transit. An ADS does not eliminate the responsibility of the user to employ the service correctly and appropriately.

Economy

AVs (and other ADS-equipped vehicles) have the potential to deliver significant positive changes to the economy through an improvement in overall accessibility—in other words the ease with which people can move from one part of a city to another. This includes easier access to job and education opportunities. Easier access to healthcare will also have an indirect impact on the economy by reducing productivity lost through illness. AVs are also expected to reduce the cost of moving goods from origin to destination. Replacing the labor cost of a driver should significantly decrease the cost of goods movement. This will be reinforced by more efficient traffic movement, which in turn will reduce the congestion experienced by goods and transit vehicles, hence increasing travel time reliability and increasing travel time reliability while reducing overall trip time.

This is expected to occur across all modes of transportation, causing a significant reduction in congestion and an improvement in trip time reliability that unlocks additional time savings. The design and production of AVs will also have a significant impact on the economy. It is well-established that the production of automobiles has a significant impact on national and global economies. For example, according to *IndustryWeek* [4], the global automobile industry employs four million people directly and many more indirectly. According to the Center for Automotive Research [5], the auto industry is one of the most important industries in the US with a historical contribution of between 3 and 3.5% of the overall gross domestic product.

It is expected that the advent of ADS technology will increase this contribution as the electronic and software content in vehicles increases. It could also be argued that

AVs will reduce the total number of vehicles by enabling more effective car sharing and multimodal trip chaining, but this is likely to be offset by an increase in vehicle miles traveled requiring more frequent vehicle replacement.

AVs will have an impact on the cost of transit services by eliminating operator costs. With respect to freight, local deliveries using AVs have the potential to revolutionize the logistics associated with this operation. However, it may be necessary to manage such deliveries to a higher level in order to avoid automated delivery congestion. This may require the application of advanced logistics and fleet management solutions.

This aspect will also require an economic analysis of ADS applications including private car, "first mile, last mile shuttles," transit vehicles, and freight vehicles. The economics will be affected by the need to provide specific vehicles for specific trip purposes. There could be unexpected effects on the local economy as human drivers are replaced by ADS technology for transit vehicles. If an economically feasible business model is to be defined and implemented for ADS technology, then consideration must be given to the operational concept, costs, and value delivered. It will be vital to ensure that there is a financially sustainable ADS business model.

Efficiency

AVs are expected to increase the efficiency of the transportation process, including efficiency gains for private automobiles, goods vehicles, and transit vehicles. Efficiency can be measured from several different perspectives, including energy efficiency, greenhouse gas emissions, and time taken to travel. It can also include reductions in the cost of accidents and improve responsiveness to incidents and events. AVs will increase efficiency across all modes of transportation by providing better management of stops and delays and smoother traffic that will reduce acceleration and deceleration, the primary cause of emissions and energy consumption. Transportation efficiency has a much wider reach than just system users.

The delivery of mobility and accessibility in an efficient manner across cities is a key factor in attracting industries and innovation to cities and is also a major influence in quality of life. In the US alone, congestion cost $305 billion in 2017 [6]. With respect to transit vehicles operating on fixed routes, efficiency gains may be offset by the need to manage trips in both directions. If the return trip has no passengers, then efficiency will be reduced. Efficiency could also be impacted by the need for different specialist vehicles for different purposes. This could also involve different branding from the private sector, with different vehicle characteristics, which will need to be managed in relation to demand for trips. AVs may not reduce the overall vehicle miles traveled due to one-way trips and the need for special vehicles for specific trip purposes. Matching supply and demand is also an important dimension of efficiency. This requires a detailed understanding of current transportation operating conditions and variations in the demand for transportation.

Energy

The energy profile for transportation services is likely to be significantly altered by the introduction of AVs. Combined with the introduction of electric vehicles, AVs will lead to better fleet management for goods vehicles and transit vehicles. With respect to the private automobile, energy impacts would be expected in terms of a reduction in vehicle hours traveled and smoother traffic flows featuring lower levels of acceleration and deceleration. They should also have an impact on the cost of maintenance.

Environment

Like energy and efficiency, the environment will benefit from the introduction of ADS technology. Results will be obtained across a wide spectrum of activities including better operation of individual vehicles, more efficient fleet management, and better overall coordination and alignment between different modes. This will include better connectivity between services, a higher level of adaptability for each service, and for comprehensive service coordination. Two-way connectivity between roadside infrastructure and AVs will also enable better advice to be provided to drivers. This will include information on environmental impact and support a comprehensive assessment of the overall environmental impact of transportation services within cities. This policy aspect also includes consideration of sustainability. Will ADS technology support a decarbonized approach to transportation as part of a wider sustainability framework for transportation? Another consideration is that AVs, along with the switch to electric vehicles, will result in different kinds of waste production, requiring different waste-handling techniques.

Ethics

Ethical behavior can also be influenced by ADS technology, as the system will replace driver decision-making and judgment in many circumstances. This can include automatically obeying prevailing traffic laws, especially with respect to speed and parking. Artificial intelligence (AI) elements of ADS equipment will ensure that ethical behavior is applied to all aspects of the transportation system. This can also include the implementation of ethical laws and the detection of law violation. The ability to control AVs remotely can also have a significant impact on auto-related crime such as car theft, hijacking, and the use of vehicles for criminal activity. ADS technology supports a higher level of continuous monitoring regarding driver behavior and vehicle operating conditions.

Alternatively, it will be necessary to develop governance arrangements that ensure that this ability is not misused in terms of invasions of personal privacy. Ethics are likely to be an extremely important aspect of ADS policy. It will also be necessary to positively address any concerns that the general public has relating to privacy and anonymity. On the one hand, the availability of better data will make transportation services more effective and efficient; on the other hand, public opinion and concern about invasions of privacy could prevent the realization of the full benefits of the technology through public resistance.

Infrastructure

Infrastructure includes a combination of traditional transportation infrastructure such as roads, transit facilities, and airports—all requiring investment decisions involving asphalt, concrete, and steel. In addition, infrastructure will also incorporate roadside sensors and telecommunications to support connected and AV. Public policy should ensure a close alignment between ADS technology and infrastructure design. Infrastructure also includes signs and markings required to guide and instruct drivers, and ultimately AVs. Infrastructure aspects of policy should also address the roles and responsibilities of the public and the private sectors. Does the public sector provide the infrastructure, while the private sector provides vehicles, or is there some other more appropriate assignment of roles and responsibilities? There will also be a need to address the interaction between vehicles and infrastructure by considering the type of support infrastructure required to enable connectivity between the vehicle and the road. Any policies for infrastructure should also take account of vulnerable road users.

Insurance

Insurance is a significant cost component in operating transportation services. Insurance companies are already experimenting with usage-based insurance, which includes the replacement of conventional underwriting tables based on age, geographic location, and vehicle type. These are replaced with other factors that take account of operating conditions and driver behavior. It is expected that this trend will continue with the introduction of ADS technology. Policy aspects related to insurance include the determination of liability for accidents. If a self-driving vehicle has an accident, who is responsible-the driver, the automobile manufacturer, or the original equipment manufacturer (OEM) that provided the ADS?

From another perspective, it could well be the case that drivers who choose to continue to operate their vehicles manually may incur an insurance penalty as the relationship between ADS technology and increased safety is understood and established. An ADS generates a considerable amount of data regarding the vehicle and driver performance—it is likely that this data will be used in the future to create a much better and more detailed understanding of the risks that are covered by insurance. This should lead to better ways to customize insurance to fit individual needs, and hopefully lower insurance rates that reflect the more detailed understanding of risks.

Land Use and Urban Planning

This can cover the definition of where AVs are permitted to operate and where they are not. It can also involve the use of vehicle data from connected vehicles and AVs to develop a better understanding of the relationship between land use and transportation demand. For example, current land use planning techniques that determine the impact of particular land uses on transportation draw on historical data based on the transportation impact of prior development. ADS data should enable a scientific approach to this, based on observed data. This policy aspect could also address one of the likely major shifts in use caused by AVs. It is possible that the use of parking lots and parking structures will be significantly impacted by the introduction of AVs that don't require to be parked.

Liability

In the same vein as the policy issues related to insurance, liability is also an important consideration. While the liability situation with respect to manually driven vehicles is well-established and understood, there will be a need to develop new policies related to liability for AVs. Liability for accidents and incidents could also be affected by new data streams available from the vehicle, which will provide greater insight into the nature and causes of such accidents and incidents.

Mobility, Connectivity, and Accessibility

Mobility, connectivity, and accessibility are all measures of the performance of transportation services within an urban area. Mobility characterizes the number of trips that can be made based on the availability of transportation services and the cost of such services. Accessibility characterizes the ease or difficulty of traveling from one part of a city to another. Connectivity is a measure of how well various transportation services are aligned and coordinated and also how well each component of the overall system can work together.

Typically, ADS policy objectives will have mobility-, connectivity-, or accessibility-related goals. This addresses the reasons for enabling and supporting ADS technology in the first place. This technology will enable the very young and the very old to have better access to transportation options, including private cars, ridesharing, and "first mile, last mile" shuttles.

ADS technology is expected to improve connectivity by supporting first mile, last mile connections from trip origin to a mainstream transportation service, and from that service to the ultimate destination. Accessibility is expected to be improved through the availability of more transportation options between various origins and destinations around a city, as well as an improved awareness of the transportation options for the user at a specific time and for a specific journey. This policy aspect will also address the use of "mobility-as-a-service" (MaaS) and automated driving techniques that support shared trips as a way of managing vehicle miles traveled.

Mobility, connectivity, and accessibility aspects will affect freight, logistics deliveries, and transit as well as private vehicles.

Modes

A balanced policy approach must take account of all modes of transportation:

- Private car
- Bus
- Rail
- Freight
- Transportation network companies
- Nonmotorized transportation (pedestrian, cycling, scooter, etc.)
- Rideshare companies (bicycles, scooters, etc.)
- First mile, last mile shuttles
- Animal powered
- Air
- Water

It is likely that ADS technology will play a significant role in providing connectivity between individual modes supporting the operation of the entire transportation network as a single system, something that has been difficult to achieve up until now. Each mode of transportation has specific needs and characteristics that must be incorporated into a equitable policy.

Operational

Policy will have several areas of impact with respect to ADS operations. Operations is one activity within an overall spectrum of activities required to support transportation service delivery. Other activities include planning, design, delivery (or "building"), and maintenance (Figure 3).

For example, policy will define areas of the transportation network in which it is permissible to operate AVs, and those areas in which it is not. Operational management will also have a considerable impact on the quality of transportation services delivered including safety, efficiency, and user experience. Operational ADS constraints could also have the effect of stifling the full delivery of these benefits. It will be important to define policy regarding the operation of transit vehicles. For example, the conditions under which control of the vehicle should be recovered from the ADS to the driver or operator will have to be defined.

AV operation on a fleet basis also requires consideration of arrangements to maintain levels of service quality. For example, under the auspices of current operational models, rental cars are cleaned and maintained at the end of every hire. Transit vehicles are also cleaned and maintained on a regular basis, with cleaning taking place every evening. With respect to transit, the operator's role extends beyond driving to include monitoring of vehicle condition and enforcing compliance to fare regulations. This can include response to unexpected spillage and customer service needs. Transit operators also have responsibility for maintaining safety and security of passengers. This will require consideration of how these roles will be addressed by ADS operation. This raises the question of AV operation or centralized fleet control.

A centralized operational model for fleet management may be able to address some of these issues. Transit operations have some particular challenges relating to social equity and a legal requirement to make transit accessible, regardless of physical ability. For example, a person in a wheelchair would typically rely on the operator to aid with boarding and securing the wheelchair. This option would not be available

FIGURE 3. **Transportation areas of activity.**

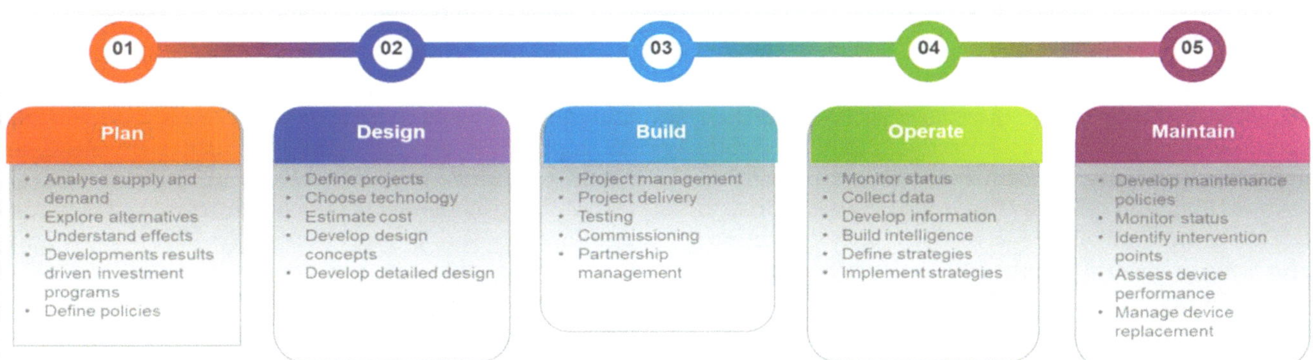

from an AV, unless an attendant was present. The needs of individuals with disabilities with respect to AVs have been covered in detail by others [7].

Partnership

Partnership aspects of policy include support for establishing and managing public-private collaborations. This involves the definition of an overall framework within which ADS implementation can be made by aligning efforts from both the public and private sectors. Alignment across the public sector is important as the transportation network on which AVs will operate falls under different jurisdictions at city, county, state, and national levels. Alignment of goals and objectives at each of these levels is a key enabler of partnerships that promote rapid, yet safe, AV introduction. Of particular concern is the need for consistency across each of these levels, avoiding conflicting policy objectives at different levels.

This could be a challenge if AV introduction is conducted using a bottom-up approach in which different local jurisdictions are involved in pilot and demonstration projects. This is currently the case in the US, where more than 1,400 individual AV trials are underway. While such focus and fragmentation may be necessary to make rapid progress and assimilate practical knowledge, it will also be necessary to apply a consistent framework to optimize the use of invested resources. With respect to public-private partnership, there is much to be gained in establishing policies that fully support the achievement of public sector objectives, with the full involvement of private sector resources.

Previous implementation models involved a clear lead from the public sector in terms of policy establishment, investment, implementation, and operations. Over a period of time, the private sector has become more involved. For example, the privatization of railways in certain parts of the world have required coordination between policies related to infrastructure and those related to the operation of the vehicles. In many cases, the private sector operates vehicles, while the public sector operates the infrastructure. In the case of ADS technology and AVs, there is a much more dominant private sector lead with respect to planning, implementation, and operation.

While there is a public sector role in providing support infrastructure and ensuring that the infrastructure features are aligned with the needs of AVs, the primary investment activity is being supported by the private sector. Since the introduction of the private car, there has been a clear division between public and private sector activity, with the public sector providing the road infrastructure and the private sector providing the vehicles. This is also the case for AVs, which has a focus on vehicle technology. Out of necessity, the private sector must take the lead in the development of technologies destined for use on vehicles.

This leads directly to a situation in which the investments of the private sector will realize public sector benefits and achieve public sector policy objectives. This places a strong emphasis on the development of appropriate partnerships to ensure that the motivation and direction of both public and private sectors are aligned as well as possible. Partnership aspects of policy must be clearly defined and must communicate the relative responsibilities and objectives for each sector.

Procurement

Procurement aspects of ADS policy also build on the need for partnership. When transportation technologies emerged, there was a clear lead from the public sector. The business model for procurement could be described quite simply as "we buy it, we own it, we control it." The need for closer partnership between the public and private sectors means that this procurement will not work in the new circumstances and environment. There is a great need for advocacy, influence, and communication to ensure that public sector objectives will be achieved by private sector initiatives. A new procurement model that incorporates partnership, advocacy, and influence in place of direction and control will be required. The two models are illustrated in Figure 4. Note that there are several variations in public-private partnership, ranging from public sector funding of private sector operations, to completely privatized self-funded operations by the private sector.

The transition from one procurement model to another can be a challenge for public agencies as procurement arrangements and organizations are built around the "buy, own, control" model. This model has the advantage of clear definition of boundaries and protection of public sector objectives but can exclude private sector resources and motivations. In addition to the procurement model, the rules and regulations governing procurement may need to be revisited in order to enable a new business model. Public agencies are constrained by rules pertaining to competitive acquisition and acceptance of lowest bid.

Public Health

Public health aspects of ADS policy can include measures to protect the health and safety of the traveling public. Such policy aspects can include the definition of sanitation and cleaning requirements for vehicles across all modes of transportation. While the private automobile is the responsibility of the owner, there is a future scenario in which an AV is provided on an on-demand basis to travelers. This would be similar to the current rental car, rideshare, or taxi model. The removal of the driver from such AVs could also have a positive impact on public health by avoiding the need for contact between operator and passengers. Of course, this would not address potential contact between passengers, unless the AV was designed for a small number of people, more akin to a private car service than a transit

FIGURE 4. **Example procurement models.**

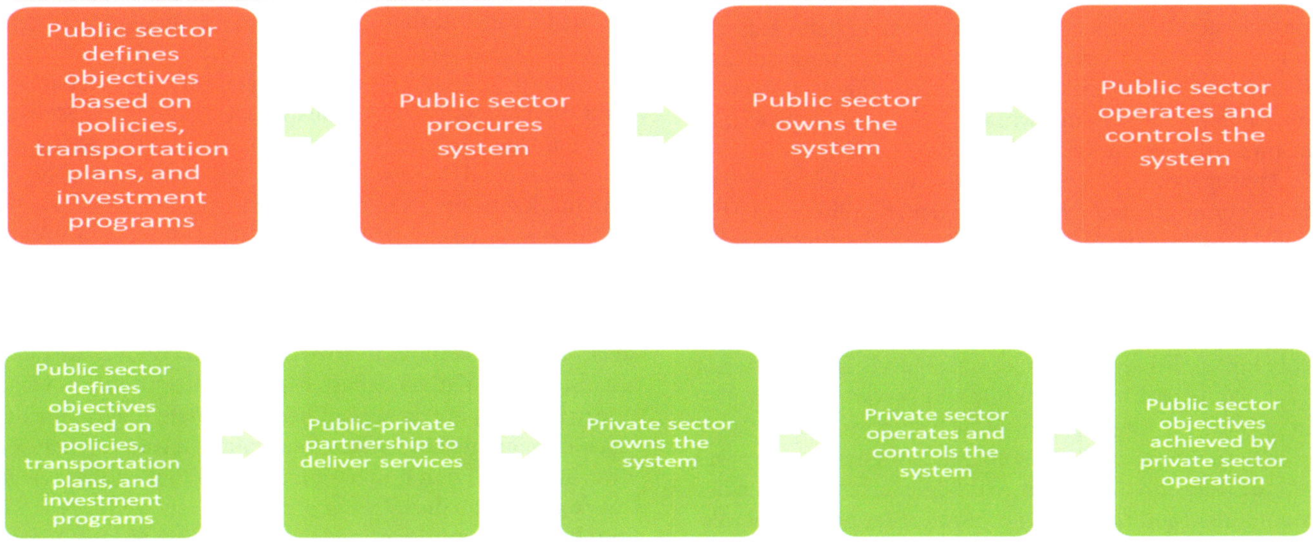

service. Public health could also be impacted by new ADS abilities to connect those in need of medical services with those services.

Regulation

Regulation is more than a policy aspect with respect to ADS technology. Regulation can be a tool to ensure the successful application of predefined policies. However, there is a policy aspect of regulation in achieving the balance between public sector objectives and the enablement and promotion of innovation and transportation technology. Like most developments in transportation, there is a need to achieve desirable effects, while avoiding undesirable side effects, or consequences. This is manifested in current transportation network operation, especially with respect to private vehicles where vehicle operating speed and vehicle parking are subject to regulations. Vehicle operations are also subject to regulations including specification of the side of the road on which to drive and appropriate behavior at stop signs and traffic signals.

Safety

One of the primary drivers behind ADS and AV introduction is the desire to achieve higher levels of safety, while still delivering efficiency and user experience gains. The operation of a transportation system of any kind requires a balance between safety, efficiency, and user experience. While safety is extremely important, it cannot be solely addressed to the detriment of the other benefits. For example, the most direct way to achieve the safest operation of a highway facility would simply be to close it. This would guarantee no accidents or fatalities but would not deliver the desired benefits of increased efficiency and improved user experience. This balance between safety and the attainment of other objectives is perhaps best exemplified in transportation by referring to the aviation industry. This industry has achieved extremely high levels of safety, while still accommodating technology innovation leading to improvements in efficiency and enhanced user experience. One of the factors in this ability to balance could lie in the operation of vehicles exclusively by highly trained specialist staff. The removal of the human driver from AV could be a similar enabler.

Therefore, safety must be considered within a balanced framework that requires the optimization of safety, along with the other policy objectives. AV safety is of prime importance, but it must support the introduction and acceptance within a reasonable timeframe. The technology adoption cycle for other technologies are indicating that it can be very effective to introduce technologies on an incremental basis under the auspices of which technologies are applied in practice, without complete development. This enables the early attainment of safety, efficiency, and user experience benefits, while practical experience and knowledge gain continues. Due diligence is required to ensure that unnecessary risks are avoided, while progress is made in applying the technology to real-life situations.

An opportunity cost approach to technology introduction would avoid waiting for technology to be perfect while incurring the loss of opportunities to save lives through the application of evolving technology. Another concern with respect to safety is a perception that ADS technology must be perfect when, in fact, a reasonable goal would be for ADS technology to be better than a human. The old adage "perfection is the enemy of the good" comes to mind.

Security

From a systems engineering viewpoint, safety and security can be viewed as different dimensions of the same subject. For example, safety could be considered "the protection of the user from the system," while security can be viewed as "the protection of the system from the user."

Appropriate policies for ADS security will be essential. The telecommunications link between the vehicle and the roadside is a new development that will attract undesirable attention as it grows in prominence and importance. The possibility that the vehicle could be controlled remotely by malevolent interests makes it essential to define arrangements for system security. Acceptable levels of security can be defined as part of both public and private sector policy that governs the development of such systems. As ADS and AV introduction places a new emphasis on data centers and communication networks, it will be necessary to consider system vulnerabilities and communication network weaknesses as part of this policy aspect. New operational models enabled by the new technology require that we revisit security arrangements.

Social Equity

Social equity relates to the creation of a consistent and even impact across the population and is of particular concern to the public sector, in terms of future of public investment. Policies should be designed to be fair with an even impact across different demographics, races, and sectors of the population. At the very least, policies should ensure that smaller groups of the population are not disadvantaged to the benefit of the larger groups.

This is an area where the public and private sectors could be in conflict. Typically, private sector motivation is driven by an economic focus to deliver to a particular target (paying) audience or segment of the population. There is also an argument that generating benefits for a specific segment of the population will have knock-on effects for the others.

ADS policies should take account of the difference in motivations and attempt to provide an equilibrium between both. While there is a public duty to preserve social equity, this needs to be balanced against encouraging the private sector to invest and take the risks required to create the market. This may involve public sector investment to subsidize services that are not economic, in order to support social equity. It has already been established that the application of advanced technology to transportation involves business models that are suitable for the public sector to lead, and those that are appropriate for the private sector to lead.

For example, traffic signals have no established mechanism for payment by the individual, therefore it is more appropriate for the signals to be implemented and operated as a public sector initiative funded by taxation. This does not mean that there is no role for the private sector in traffic signal operations. It is feasible that the private sector can operate traffic signals on behalf of the public sector. In this case, it is likely that the funding required will come from the public sector, and that operational objectives are set by the public sector as well.

This policy aspect has a close relationship with partnership aspects discussed previously, as it is essential to establish and manage effective partnerships between the public and private sectors if a reasonable trade-off between social equity and market factors is to be achieved.

Standards

It is not clear if standards drive policy or if policy drives standards. A typical market-based approach to standards would mean that they are created at the appropriate stage in the lifecycle development of a technology market. From this perspective, standards can be viewed as a tool that enables larger markets through the introduction of standardized approaches and hence economies of scale. Another way to look at standards is in support of public procurement of technology to ensure a complete understanding of what is being procured and to make sure that the public sector gets what it purchases. Standards also help in the development of bidding documents and specifications to support public sector procurement of technology-based products and services.

To some extent, ADS policies will have an influence on standards by describing and defining what the public sector wishes to achieve. Well-informed policy will also take account of existing and emerging standards. Observation of how markets for technology applications and transportation develop suggests that, at the beginning of the lifecycle, there is little demand or desire for standards to be developed. At this stage of the market, service and solution providers are motivated to protect investments in technology and preserve market share through fragmentation and focus on local needs. However, as the market grows, it becomes apparent that the creation of a larger market can be stimulated by seeking standardization and moving beyond initial fragmentation. ADS policy should take account of the dynamics of market growth, especially with respect to standardization aspects of policy.

Technology

In a similar manner to standardization, the development of technology can be seen as either influencing or reacting to policy. On the one hand, good policy will take account of the current state of available and emerging technologies. For a policy to be effective, it should build on a good understanding of the capabilities and constraints of technology. On the other hand, it can be argued that technology development is driven and influenced by policy, as these dictate market requirements.

ADS policy should consider current and emerging technology capabilities to ensure that objectives are achievable. Policy should also acknowledge the benefit of encouraging and fostering innovation to ensure that the benefits of technology are realized at the earliest possible opportunity. It is apparent that there is an inseparable link between policy and technology. While technology can evolve initially with no regard to policy, at this stage, technology could be viewed as a solution in search of a problem.

Once the technology matures significantly to form the basis for products and services, there is a need to understand and characterize the target market and customers for the products and services. It is highly likely that policy will be one of the factors considered when characterizing the market. Therefore, policy must take account of technology capability and constraints.

Telecommunications

Telecommunications aspects of ADS policy will focus on the necessary connectivity between the vehicle, other vehicles, and roadside infrastructure. Current US policy is evolving with a migration away from dedicated short-range communications (DSRC) to a more agnostic approach that includes cellular wireless. DSRC requires that transceivers are installed along the roadside and that they are capable of communicating with similar transceivers installed in vehicles. Typically, the public sector has responsibility for the investment in the appropriate roadside infrastructure, while the private sector has responsibility for the installation of the in-vehicle units. One policy aspect related to telecommunications could be a definition of investment in roadside infrastructure that aligns with parallel investment in vehicle units.

Testing

ADS policy related to testing feature a particular focus on the terms and conditions that enable open-road testing. The SAE EDGE Research Report on *Unsettled Topics Concerning the Field Testing of Automated Driving Systems* addresses this in some detail [3]. Here again, policy must balance multiple objectives: the primary one being the preservation of public safety. On the other hand, there is a need to encourage innovation and foster the development of the ADS market by allowing practical experience and knowledge gain along with public exposure to the technology. Early approaches to testing have a focus on the adjustment of the operating environment to suit technology capabilities and limitations. Perhaps a more appropriate approach would be the development of technology capabilities to suit current operating environments.

In addition to physical testing of hardware and devices, there is a need for software testing to address anomalies in coding and to identify unintended, unanticipated consequences. There are some good examples of previous practice such as NASA's Space Shuttle software that had minimum defects. Testing should also define the need for software updates and the policy implications of those updates. The software industry also provides existing models for software development using the concept of "alpha and beta" testing. In alpha testing, an isolated artificial environment is used to test the software. In beta testing, real-world environments are used with restricted access in terms of numbers of users and geographic location. Policy should define the appropriate time and place for each strategy.

Consumer research also provides existing models for testing AV user reactions with a technique known as A/B testing [8]. This involves experiments where two or more variants of a system are shown to users at random, and statistical analysis is used to determine which variation performs better for a given goal.

Initial testing activity indicates that a bottom-up approach is being taken for ADS and AV testing. Field testing activities are governed by locally derived policies that are specific to the needs of the location. While national-level advice and guidance is available, the current situation is that testing activities are being led at the local level. The initiatives underway to establish a national approach to this, particularly in the US, are spearheaded by the National Highway Traffic Safety Administration (NHTSA) [9].

Vehicle

Policy aspects related to the vehicle specifically are typically addressed at a national level. For example, NHTSA supports education, research, safety standards, and enforcement related to vehicles and highways. This includes the development and enforcement of vehicle performance standards and partnerships with state and local government to reduce fatalities, injuries, and economic losses from motor vehicle crashes [10].

Public policy can regulate AVs and ADS equipment by specifying where they can be used, what equipment is utilized, and the conditions under which the AVs will operate. Policy can also have an impact on the rate at which ADS technology is adopted by influencing things like insurance. Policy that defines liability and constrains risk can have a significant influence on market growth.

Workforce Development

Workforce development relates to the availability of a current and future workforce with appropriate skills and capabilities that support current and future industry activity. A formal definition is as follows [11]:

> *Activities, policies and programs employed by geographies to create, sustain and retain a viable workforce that can support current and future business and industry.*

ADS workforce development can have both public and private sector focus. With respect to the public sector, policies should be designed to encourage and foster the development of a workforce that is capable of fully supporting current and future ADS technology. This can involve retraining people who are currently employed as drivers or operators, as these jobs may eventually disappear from the labor force.

Policy is based on the detailed understanding of the current situation with respect to workforce deployment and an appreciation of the future needs for ADS products and services. Public policy can provide support and enablement for understanding the needs and adapting the workforce accordingly. With respect to the private sector, workforce development is of particular interest as an enabler to the development of the market, and for its impact on the ability to produce products and services.

Policy Aspect Prioritization

There are a lot of policy aspects to be considered at one time. To support a resource-leveling approach, these aspects have been categorized to define a three-stage approach. The categories are "priority," "short," and "medium" (Figure 5).

Alignment with Other Local Government Policies

ADS policies cannot be considered in a vacuum. It is essential that such policies align with other public sector strategies at the various levels of government. For example, policies related to transportation infrastructure development need to be aligned with ADS policies. Across the spectrum of policy aspects described previously, it is important to consider the wider context within which they are established and implemented. The aspects described previously can be used as a checklist to ensure that a comprehensive approach is taken to the alignment of ADS policies with other related public sector policies at various levels.

Policy alignment is essential to ensure that a systems approach is taken toward AV introduction and the operation of transportation networks as single systems. This involves the harmonization of various harmonizing efforts to avoid potential conflict and to create a cohesive reinforcing effect.

FIGURE 5. Policy aspect prioritization.

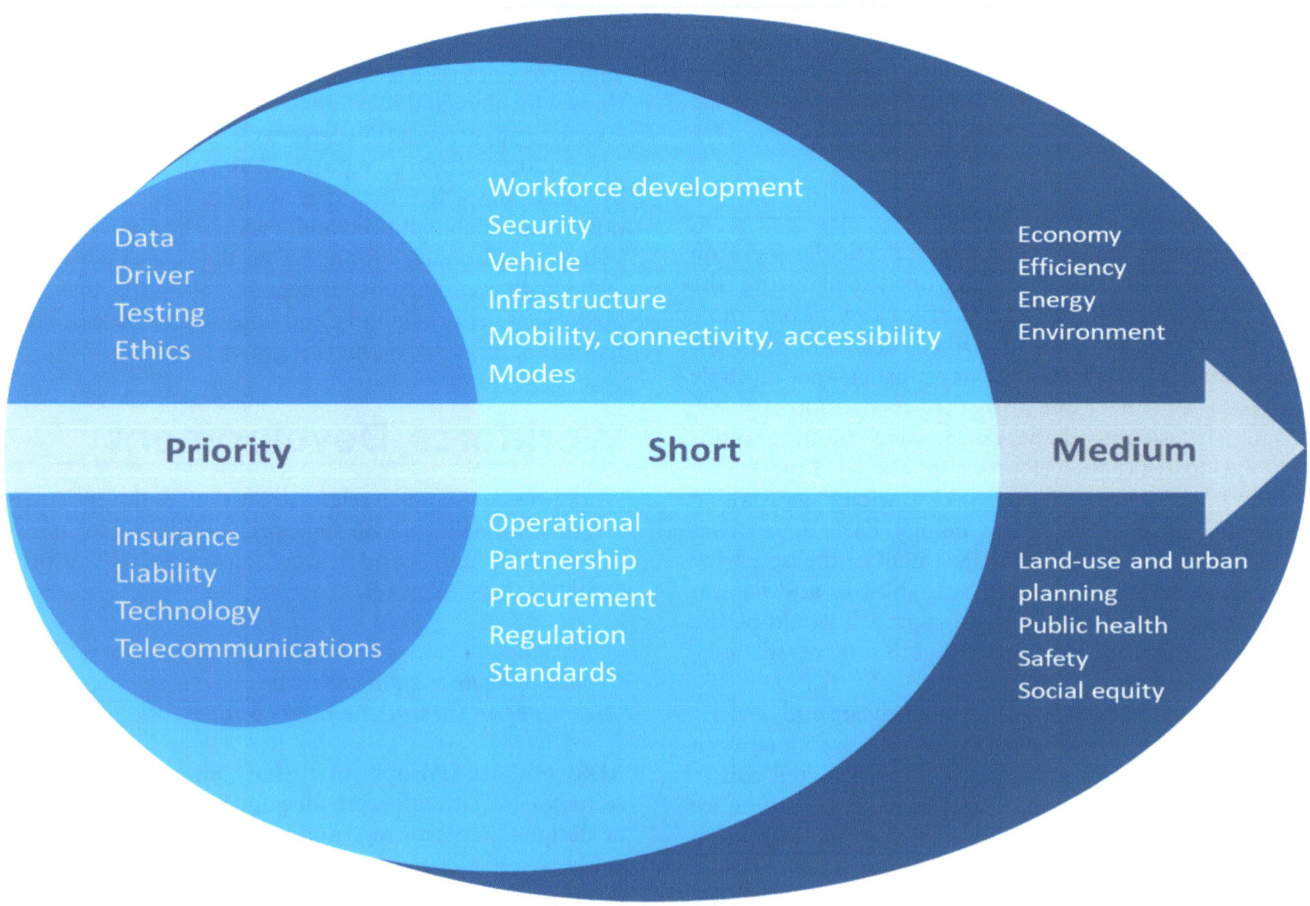

Examples of ADS policy alignment include managing transportation policy for

- Infrastructure investments
- Safety
- Economic development
- Modal use
- Demand management
- Land use

Policy alignment can also include ADS harmonization efforts across national, state, and local levels.

Stakeholders

In addition to considering ADS policy, it is useful to identify and define the stakeholder groups involved (Table 2).

TABLE 2. **Stakeholders.**

	Stakeholders
1	Academic institutions
2	Aftermarket and auto parts
3	Airlines
4	Airports
5	Auto repair companies
6	Automobile manufacturers
7	Banks and credit unions
8	City planners
9	Commercial vehicle manufacturers
10	Drivers
11	Energy and petroleum companies
12	Fast food restaurants
13	First responders
14	Hotels and motels
15	Industrial fleet operators
16	Information technology providers
17	Insurers
18	Legislators and regulators
19	Media and entertainment
20	Mobility providers
21	National and state infrastructure managers
22	Original equipment manufacturers
23	Parking lot and building owners
24	Ports
25	Real estate
26	Rental car operators
27	Ride share operators
28	Road builders and designers
29	Shipping and trucking companies
30	State and local governments
31	Taxi companies
32	Telecommunications providers
33	Transit agencies
34	Transit passengers
35	Transit vehicle manufacturers
36	Transportation network companies
37	Vehicle dealers and retail network

Academic Institutions

Academic institutions such as universities and colleges have an important role to play in the ADS development and AV implementation. These roles include research, participation in testing (including hosting AV test tracks), and delivery of the education required for workforce development. Academics will be expected to play a significant role in ADS technology development from fundamental research to practical application, and as independent arbitrators of technology capability and limitations. This is an important role in the definition of policy.

Aftermarket and Auto Parts

The vendors and manufacturers involved in the automotive aftermarket can be considered as important ADS stakeholders. While the main market for vehicle accessories and parts will continue, it is likely that this will be expanded to include ADS components and other accessories required to support the rollout of such systems.

Airlines

ADS technology could enhance the driving experience to the point where driving between two locations is a viable alternative to short-haul airline travel. Airlines may also need to widen their current view of the market to encompass MaaS integration. The air trip can be viewed as one link in an overall trip chain from origin to destination that may involve multiple modes of transportation. The availability of ADS-equipped "first mile, last mile" shuttles and other ADS-enabled ridesharing possibilities will change the context within which airline travel currently takes place. This will affect both freight and passenger services.

Airports

Similar to airlines, airports will have to adjust operating procedures to accommodate ADS technology for both freight and passenger services. This could involve changes in road geometry around airports to accommodate AV passenger drop-off and automated people movers between airport location such as terminals. Here again, airport authorities may have to change their perspective to include a wider view of transportation, with the airport as a hub within a wider network of modal travel opportunities.

Auto Repair Companies

An obvious impact on auto repair companies would be a reduction in the number of repairs necessary as ADS technology reduces accident rates. Auto repair companies will also be expected to expand current offerings to include aftermarket ADS installation, maintenance, and repair work as a part of the overall vehicle servicing market.

Automobile Manufacturers

Automobile manufacturers are already adapting to the design, development, and manufacturing of AVs. While manual driving systems will continue to proliferate in the short term, it is expected that ADS technology will replace them over a period of time. Automobile manufacturers could see their target market segments change because ADS technology enables mobility of on-demand services operated by commercial fleet companies. This could lead to a reduction in individual vehicle ownership.

A study by the University of Michigan [12] estimated that self-driving vehicles could cut average ownership rates of vehicles in the US by 43%. The study also noted that ADS technology could result in a 75% increase in individual vehicle use. Fewer cars do not necessarily equate to less miles traveled. This could be an extension of the current rental car market or transportation network companies migrating to driverless vehicles. Most of the major automobile manufacturers have responded to this by incorporating MaaS offerings into their future plans.

Banks and Credit Unions

Banks, credit unions, and other financial institutions that lend money for vehicle purchases will be affected by the availability of ADS technology. Such systems can enable changes in the vehicle ownership model away from sole ownership by an individual to fleet operations and car sharing models. Car sharing could become a prevalent ownership model, requiring financial institutions to adapt terms and conditions to enable multiple ownership models for vehicles.

City Planners

City planners use zoning and other regulations to control the shape and growth of metropolitan areas. There is a well-established relationship between transportation accessibility and desirability of land. Over time, ADS technology could modify land use and the shape of urban areas by influencing transportation accessibility. City planners will also have to cope with changes in the demand for parking as AVs are capable of "24/7" operation, with no requirement to park the vehicle.

Commercial Vehicle Manufacturers

In a similar manner to automobile manufacturers, commercial vehicle manufacturers will be required to adapt to a new ADS-inclusive market for local and long-haul freight deliveries. Vehicles are expected to have a much higher electronic content and be designed for 24/7 operation. This could result in vehicles covering a larger number of miles traveled in a shorter period. This in turn will affect maintenance cycles and warranties.

Driver

Freeing the driver from the need to control the vehicle will have a significant consequences and could allow transportation to become a secondary function that occurs while travelers are engaged in other activities. This can include working and conferencing as a vehicle is driven by the ADS, the ability for the very young and the very old to make use of private travel, and new alternatives for disabled travelers. Driving is also promoted as a fun experience that many people consider as a hobby or leisure activity. This perception may change as ADS technology obviates the need for driving.

Energy and Petroleum Companies

It is likely that ADS technology will result in more efficient use of the transportation network, reducing fuel consumption and emissions by avoiding unnecessary acceleration and deceleration. Vehicle platooning and other techniques will increase the efficiency of road use and impact the demand for energy and gasoline. This, combined with the introduction of electric vehicles, will have a significant impact on energy and petroleum companies

Fast Food Restaurants

Fast food restaurants are already being impacted by the availability of transportation network company resources to deliver food to the customer's home or office. The introduction of AVs will make this a more desirable option as the cost of the driver is removed from the process. Fast food restaurants will be able to deliver quickly and cheaply through the use of AVs.

First Responders

ADS technology could be deployed for emergency vehicles, refocusing first-responder roles to dealing with the emergency, rather than operating the vehicle. ADS implementation should also improve response times and response time reliability through better management of transportation conditions.

Hotels and Motels

Hotels and motels also include sharing services (such as Airbnb). These can be significantly impacted by the new ability for a driver to sleep in an AV while traveling. This would reduce market demand for overnight accommodations.

Industrial Fleet Operators

ADS technology has the potential to substantially impact industrial fleet operators. A significant cost of fleet operation lies in the need to train and pay drivers to operate vehicles. In the case of freight and transit vehicles, the driver is not traveling for their own purposes, but is being paid to drive in support of goods delivery or passenger travel. This will include car rental operators and freight operators. Transportation network companies can also be considered as industrial fleet operators in this context

Information Technology Providers

Technology and communications companies will build the networks for AVs to send and receive data. While it is considered that these AVs will be autonomous from a driving perspective, it is likely that some form of central fleet management will be applied to enhance the operation of the vehicles. This will enable shared information across the fleet and detection of maintenance and operational issues. Interaction between the vehicle, the back-office, and other infrastructure will depend on information technology and communications.

Telecommunication providers will also be required to accommodate higher demand for bandwidth as the data flow between vehicle and infrastructure increases. A central back-office also enables the application of AI and analytics techniques to provide further insight into operating conditions and make predictions regarding future conditions. The in-vehicle equipment will be provided by information technology companies and OEMs working on behalf of the automobile and vehicle manufacturers.

Insurers

Insurance companies can be impacted in multiple ways by the introduction of ADS technology. It is anticipated that approximately 80% of accidents will be avoided through the ADS implementation. A 2015 study estimated that AVs could reduce automobile premiums by $1,000 per annum, leading to a contraction of the auto insurance industry [13].

Legislators and Regulators

This group of stakeholders was addressed extensively in the section on policy aspects. It is expected that legislators and regulators will face new challenges and opportunities regarding ADS technology. These include the need to regulate ADS and AV testing and operations and take advantage of the new opportunities available for "first mile, last mile" automated shuttles, and a more coherent approach to the overall transportation system. This can also include regulation of the information delivered to the vehicle. These stakeholders-which include law enforcement and judicial entities-have considerable vested interests in the status quo and may face significant challenges in adapting to ADS technology.

Media and Entertainment

Media and entertainment stakeholders will mainly be impacted by the availability of new opportunities to deliver information and advertising inside the vehicle. Vehicle occupants—freed from the driving task—will likely be able to commit more attention to media consumption.

Mobility Providers

In addition to transit operators, transportation network companies, and rental fleet operators, organizations that deliver other mobility services will also be impacted by ADS technology. These include rideshare enterprises, public paratransit services, and nonemergency medical transportation. Mobility providers will have new opportunities to harness the power of ADS technology to reduce costs and make services more flexible. There could also be challenges as barriers to entry are reduced and the level of competition for mobility providers increases.

National and State Infrastructure Managers

Depending on the degree of ADS autonomy, there will be a degree of reliance on the infrastructure supporting ADS technology. This could be in the form of telecommunication devices installed at the roadside, road marks, and signs that support automated driving, and other means of improving the performance of ADS technology through infrastructure support. National and state infrastructure managers will have additional responsibilities for roadside infrastructure and harmonization of physical infrastructure with ADS technology infrastructure.

Original Equipment Manufacturers

OEMs will provide ADS technology in the form of components and complete systems. These companies may come from outside the traditional automotive industry or could be an extension of the activities of current automotive OEM companies. These enterprises will be faced with the challenge of adapting their business to new products and services, while also being presented with new opportunities for business expansion. Traditional automotive equipment will be supplemented by new products and services related to ADS technology such as hardware, software, sensors, and vehicle communication systems.

Parking Facilities and Building Owners

ADS introduction could mean a radical change in the demand for parking as ADS-equipped vehicles may operate on a 24/7 basis and do not require parking space. The vehicle can drop the traveler off, drive around until required, or be assigned to a commercial fleet operator to make money on behalf of the owner. It could also be the case that individual ownership of vehicles diminishes and is replaced by on-demand, unlimited transportation services, provided by private third parties. Here again there would be no need to park the vehicle. Similarly, building owners may be required to adapt to the new vehicle ownership model by providing less facilities for car parking and more facilities for drop-off and pickup.

Ports

Seaports will have the opportunity to enhance internal operation with ADS technology for goods staging and marshaling. This is likely to extend beyond the port as automated trucks are used to deliver goods as part of an overall multimodal trip chain. As mentioned previously, it is expected that the overall multimodal trip chain will benefit from better connectivity and communications, leading to more efficient and effective management of the entire trip chain from origin to destination.

Real Estate

An ADS impact on land use will also impact the value of land and property. There is a well-established relationship between the value of land and accessibility to good transportation. When AVs propagate into the market, it will likely lead to a change in the relative value of land and property.

Rental Car Operators

Rental car operators could benefit from new abilities to run AV fleets and deliver on-demand transportation services to consumers. This would extend current rental car operations by enabling vehicles to provide on-demand transport and then redispatch to the next customer after the trip. Automated rental cars could drop passengers off at the entrance to the airport terminal and then drive themselves back to the rental car depot or to the next customer.

Rideshare Operators

Transportation network companies and rideshare operators who provide on-demand access to electric bicycles and scooters could be impacted by new transportation possibilities afforded by ADS technology. These can include "first mile, last mile" automated shuttles, which could reduce the demand for rideshare for electric bicycles and scooters, or alternatively stimulate the demand for such services by creating multimodal trip chain opportunities. Rideshare operators may also extend or adapt their businesses to accommodate AVs. There is potential for such organizations to adopt a MaaS approach, providing travelers with multiple transportation options for specific journeys, enabling a central reservation to be made for all links in the trip chain, and a single electronic payment.

Road Builders and Designers

Ultimately, ADS technology could replace the need for barriers, road marks, and signs. This would have a significant impact on the way roads are currently designed, constructed, and operated. It is feasible that running lanes on highways could be narrower as ADS technology delivers a higher degree of lateral control. The reduction of accident and breakdown potential could also see the removal of hard shoulder facilities. Future roads and highways could also feature integrated interchange points to facilitate transfer from one mode of transportation to another. Such mobility hubs are already under development in several parts of the world, with new major highway project designs featuring such interchange points.

Shipping and Trucking Companies

With respect to shipping and trucking companies, the obvious and largest impact will be in the removal of the driver, hence reducing the cost of operations. ADS technology could also lead to more effective itinerary planning and scheduling and allow for 24/7 operations that can extend the current reach of these companies. This will further reduce the cost of operations, along with the reduced cost of accidents and repairs related to enhanced levels of ADS safety.

State and Local Governments

State and local governments with responsibility for transportation service delivery can be significantly affected by the introduction of AVs. Policies and proposed investment programs will be required to adapt to and align with new ADS capabilities. This will include changes in land use, new interchange possibilities, and new abilities to provide comprehensive "first mile, last mile" services using a combination of ADS technologies and conventional transportation. This will also require a new approach to policy development and establishment of effective public-private partnerships.

Taxi Companies

Taxi companies stand to benefit from the cost reductions associated with removing the driver from the vehicle. While ADS technology may lead to an increase in vehicle cost, this should be more than compensated for by the reduction in labor cost. Taxi companies can also benefit from the ability to operate on a 24/7 basis and avoid any shortage of professional drivers.

Transit Agencies

Transit agencies, already challenged by changes in the target market as user preferences shift, will be faced with additional demands in the form of competition from other mobility providers employing AVs. They will also be provided with opportunities to harness ADS technology to make transit services more effective and more efficient. This can include the operation of "first mile, last mile" automated shuttles and the automation of entire vehicle fleets, removing the cost of labor associated with operators. Additionally, 24/7 operation is a potential benefit for transit agencies.

Transit Passengers

ADS technology will require transit passengers to adapt to new interfaces, requiring interaction with the machine instead of a human (if drivers and/or operators are eliminated). Transit passengers should also benefit from improved reliability and travel times, as well as cost reductions associated with operational efficiencies. New service possibilities could also provide passengers with an enhanced range of services—including comprehensive door-to-door MaaS services.

Transit Vehicle Manufacturers

Like other automotive OEMs, transit vehicle manufacturers will face both challenges and opportunities related to incorporating ADS technology into their current products. The challenges will come from the need to adapt to new technologies and incorporate these into their existing manufacturing operations. ADS technology will also lead to an expansion of product offerings from transit vehicle manufacturers. This could also include a widening of the range of transit vehicle sizes to provide custom vehicles for different trip purposes.

Transportation Network Companies

Transportation network companies are already experimenting with ADS technology. It is expected that this will continue into the future and lead to a situation where the driver is replaced by an ADS. This should reduce the operating cost and increase the flexibility of services. It should also reduce cost of insurance and enable 24/7 operation.

Vehicle Dealers and the Retail Network

Vehicle dealers and the retail network that sells consumers vehicles will need to adapt to sales and marketing of new products based around ADS technology. This will also extend to the maintenance and repair of AVs in addition to the current manually driven vehicles. One interesting idea is that an AV could drive itself to a service provider for repair or maintenance, requiring changes in traditional facilities, organization, and processes.

Characterizing Stakeholder Interest

ADS stakeholder interest will vary according to individual needs, with the assumption that those needs will be related to positioning of the stakeholder along the evolution of interest process (Figure 6). Four stages of evolution of interest have been identified:

- **Awareness:** The stakeholder has a need for a basic awareness of ADS capabilities and their impact on the stakeholder group.
- **Insight:** The stakeholder has a need for more than basic ADS awareness and is interested in developing some insight into ADS effects and possibilities.

FIGURE 6. Evolution of interest.

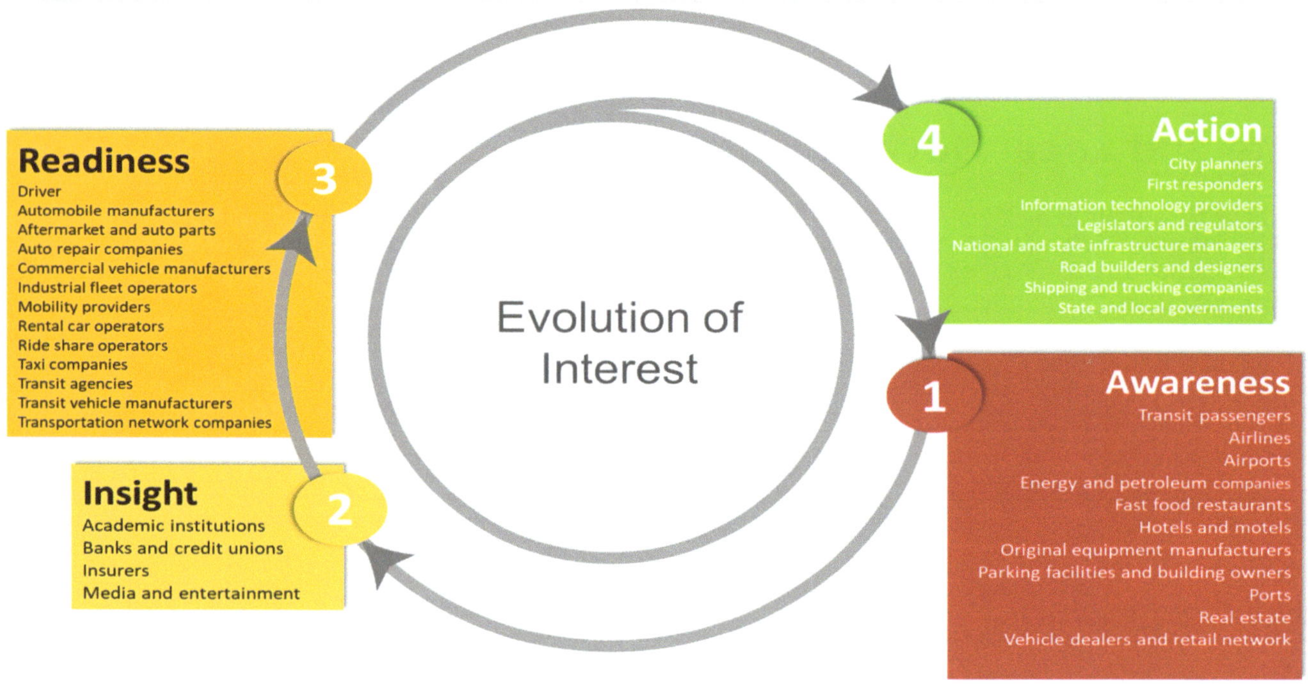

- **Readiness:** The stakeholder is at an advanced stage of researching ADS technology and requires information as part of a readiness program.
- **Action:** The stakeholder is ready to act based on insight and information regarding ADS technology.

Unsettled Issues Regarding Policy Aspects of ADS

Previous sections of the report have provided information regarding ADS architecture and taxonomy, policy aspects, and stakeholders. From this body of information, a number of unsettled issues can be identified. These are issues, challenges, questions, or opportunities where further research and practical experience are required to provide a definitive answer. Table 3 provides a summary of unsettled issues that have been identified in the research leading to the development of this report.

1. **Policy consistency**
   A lack of consistency exists between the different levels of government—local, state, and national—with respect to the ADS policy. Considerable progress has been made very rapidly by focusing efforts at the local level, leading to local policy determination and application that is not consistent at the national level. It is reasonable to assume that this lack of consistency also applies to alignment of transportation policy related to AVs and those more general policies already in force. Harmonization across government levels and across different policy categories has yet to occur.

TABLE 3. Unsettled issues.

	Unsettled issues
1	Policy consistency
2	Effective private sector engagement
3	How to achieve multimodal transportation policy objectives using AVs
4	Incomplete awareness of ADS and AV impacts
5	Lack of available standards
6	Lack of public policies on regarding data sharing and use
7	Lack of understanding regarding public sector role in ADS adoption
8	Lack of understanding regarding AV land use impacts
9	Policy communications to the general public
10	Procurement policy focused on an old model
11	Risk management and liability issues are unclear
12	Lack of understanding regarding ADS impacts to workforce development

2. **Effective private sector engagement**
It could be argued that the private sector has been effectively engaged as most of the initiatives involving ADS and AV pilot testing are led by the private sector. However, there is a lack of engagement between the public and private sector to ensure that objectives and motivations are coordinated and harmonized. Effective private sector engagement would foster innovation and allow the rapid establishment of an ADS market, while preserving public sector objectives in terms of safety, efficiency, and user experience. Public sector objectives that have been defined for the overall transportation network should also be aligned with the various levels of public and private sectors activities. It is not yet clear how they can work together effectively to achieve these objectives and optimize ADS introduction.

3. **How to achieve multimodal transportation policy objectives using AVs**
AVs, by their very nature, focus on just one modal aspect of the overall transportation network—the vehicle and highway combination. The public sector faces a wider challenge to ensure that the most efficient mode of transportation is taken for each specific journey, enabling balance between the demand for various modes and the supply. It is not yet clear how multimodal transportation policy objectives will be achieved using ADS technology.

4. **Incomplete awareness of ADS and AV impacts**
While there is significant activity associated with ADS and AV pilot testing, there is currently insufficient practical experience to support a complete understanding of potential ADS and AV impacts on the transportation network and on the traveler. There is also a lack of information sharing on practical lessons learned as most of the testing activity is led by the private sector (which considers such information a proprietary business differentiator). A complete awareness of the impact is an essential ingredient in communicating the values and benefits to the general public.

5. **Lack of available standards**
While considerable progress has been made in the development of appropriate standards, there is not yet a complete set of standards to support widespread ADS introduction. In the US, the federal government has adopted a stance that ADS standards are voluntary and there is not yet a need for specific regulation. This includes a lack of state and federal regulation for ADS and AV testing on public roads.

6. **Lack of public policies regarding data sharing and use**
While this issue encompasses a wider field than ADS technology, the lack of public policies on data sharing and use could be a hindrance to the rapid deployment of AVs. Such systems are likely to generate a large volume of data that can be beneficial to transportation operations in terms of improving safety, increasing efficiency, and enhancing the vehicle user experience. The data can also play a crucial role in accelerating the ADS implementation through knowledge and information sharing. Public policy on data sharing and the agreements required to support effective data sharing, especially with respect to ADS technology, are still a developing area.

7. **Lack of understanding regarding public sector role in ADS adoption**
There is a lack of public sector understanding and awareness with respect to potential roles in ADS implementation. Ways to communicate the characteristics and benefits of ADS technology have yet to be developed on a consistent basis. ADS information is generated at the local level as a product of pilot testing. While there are some sources of information at the national level, the overall ADS "narrative" is somewhat fragmented and incomplete

8. **Lack of understanding regarding AV land use impacts**
AVs have the potential to make significant impacts on land use. Such systems will significantly change service availability, accessibility, and mobility in urban areas. It's an established fact that there is a relationship between transportation and land value. The broader AV land use impacts are not yet completely understood, and very little discussion on the subject has occurred.

9. **Policy communications to the general public**
Communicating public and private sector policies to the general public remains a challenge. The lack of consistent national message regarding the characteristics, nature, and benefits of ADS technology remains a challenge in this respect. There is little in the way of guidance on how best to approach policy communications to the general public.

10. **Procurement policy focused on an old model**
Procurement policy and arrangements are typically focused on existing models that can be characterized as "buy it, own it, control it." With the presence of significant investment and testing from the private sector, it is essential to support new procurement models with policies that support public-private partnership. These models should involve influence, advocacy, and communication from the public sector to the private sector. It is highly unlikely that public sector investments alone will lead to the widespread ADS implementation.

11. **Risk management and liability issues are unclear**
Due to lack of practical experience in AV operation on public roads, risk management and liability issues are still unclear and unsettled. A catalog of the associated risks and their severity and likelihood is not available. A clear picture has not yet been established in the relative roles and responsibilities of the various

stakeholders from a legal perspective. Liability has yet to be analyzed and communicated.

12. **Lack of understanding regarding ADS impacts to workforce development**
At a high level, there is some insight emerging on how AVs will affect workforce development. There is a significant need to define the skill set required for the support of current and future ADS technology and to define the educational requirements to support future growth. Currently, there is no detailed understanding on the relationship between ADS technology and workforce development.

Outlook

The future looks bright for ADS technology. There is a considerable amount of private sector investment and public sector interest in supporting the application of technology to public roads. The development of ADS technology also seems to be at the leading edge of the application of AI to transportation industry. A recent report by the Brookings Institute [14] estimates that more than $80 billion has been invested globally.

While there are significant challenges to be overcome in moving toward wide-scale ADS deployment, the level of activity and investment suggests that there is considerable optimism with respect to the future. National governments have made ADS development and introduction a priority and these governments are starting to converge with the private sector to deliver practical, open-road results. To underwrite the chances of success, it will be necessary to address some of the challenges and develop appropriate means to address them in a balanced and effective manner.

Looking back over the past 20 years, it is obvious that considerable progress has been made recently toward the realization of ADS technology and the wide-scale deployment of such systems is closer than ever.

Summary

The research conducted indicates that there is a very high level of interest in the successful ADS technology application around the globe. While initiatives are being led by the private sector, there is considerable public sector involvement and cooperation, and the pilot introduction stage is currently underway. There are, however, several policy challenges still to be addressed to clear the way for successful and wide-scale deployment.

Policy challenges relate to situation where technology capability has moved ahead of policy development. Twenty-seven separate policy aspects in which work is still required have been identified. These include policies related to data sharing and usage, economic impact, infrastructure investments, insurance and liability, procurement, safety, standards, and workforce development, among others. In order to consolidate the rapid progress that has been made, it will be necessary to address the full range of policy challenges through the development of appropriate policies that balance public sector objectives, with the stimulation and encouragement of innovation from the private sector.

SAE EDGE™ Research Reports

SAE EDGE™ Research Reports, like the present report on "Unsettled Issues Regarding Policy Aspects of Automated Driving Systems," are intended to push further out into still unsettled areas of technology of interest to the mobility industry. SAE launches these reports before attempting to form a joint working group, let alone a cooperative research program or a standards committee.

SAE EDGE™ reports are intended to be quick, concise overviews of major unsettled areas where vital new technologies are emerging. An unsettled area is characterized more by confusion and controversy than established order. Early practitioners must confront an absence of agreement. Their challenge is often not to seize the high ground but to find common ground. These scouting reports from the frontiers of investigation are intended merely to begin the process of sorting through critical issues, contributing to a better understanding of key problems, and providing helpful suggestions about possible next steps and avenues of investigation.

SAE EDGE™ Research Reports, therefore, are fundamentally distinct from the more formal working groups approach and far removed from the more mature research program and standard's development process.

Next Steps for ADS Policy

This publication should be considered only as a first step toward clarifying the issues regarding ADS policy. The intention behind this and other SAE EDGE™ Research Reports is to start a dialogue among interested parties on important industry-wide topics that require further attention. The expectation is that these explorations of unsettled areas of technology will lead to the formation of working groups and, ultimately, committees that can address and resolve the issues they raise, producing a framework for developing a common vocabulary of definitions, best practices, protocols, and standards needed to support continued progress toward safer and more innovative products.

The experts' meeting that gave rise to this publication demonstrated a great willingness on the part of the industry to define the terminology, procedures, and eventually the standards needed to enable ADS policy development to move ahead as quickly and efficiently as possible.

TABLE 4. Next steps for ADS Policy.

	Aspect	Next steps
1	Data	Develop guidance on data sharing and agreements required to support ADS data use
2	Driver	Converge existing policies regarding traditional transportation with ADS policies
3	Economy	Develop a better understanding of the economic impacts of ADS technology
4	Efficiency	Understand the relationship between ADS technology and transportation efficiency
5	Energy	Understand the energy impacts of ADS technology
6	Environment	Define the impact of ADS technology and connectivity on the environment
7	Ethics	Develop a code of ethics for both the public and private sectors
8	Infrastructure	Synchronize infrastructure investments with AV deployment
9	Insurance	Define a detailed assessment of the impact of ADS technology on insurance rates
10	Land use and urban planning	Define the opportunities and challenges associated land use, urban planning, and AVs
11	Liability	Conduct an analysis of liability associated with AVs
12	Mobility, connectivity, accessibility	Assess the effects of AVs on mobility, connectivity, and accessibility
13	Modes	Define the modal impact of AVs
14	Operational	Assess the operational impact of ADS technology on the transportation network
15	Partnership	Develop guidance for effective public-private partnership establishment and management
16	Procurement	Review procurement approaches and develop new models that better support effective engagement of the private sector
17	Regulation	Develop a full set of regulations and synchronize with existing traffic and transportation
18	Public health	Understand the opportunities and challenges related to public health and ADS technology
19	Safety	Develop communication tools to explain the safety impacts of AVs to the general public
20	Security	Develop guidance and protocols to ensure the security of data generated from and ingested by AVs
21	Social equity	Ensure the preservation of social equity based on a detailed understanding of the effects of AVs
22	Standards	Continue work to develop an appropriate set of standards to cover all aspects of ADS design and deployment
23	Technology	Ensure that policy and technology capability are properly synchronized
24	Telecommunications	Define clear policies that state the need for telecommunications facilities
25	Testing	Ensure the policies governing testing on public roads are consistent and comprehensive
26	Vehicle	Define policies that preserve public safety while fostering innovation
27	Workforce development	Understand the likely impact of ADS technology on workforce development

This SAE EDGE™ Research Report on "Unsettled Issues Regarding Policy Aspects of Automated Driving Systems" identifies the following key topics for further pursuit, both through continued informal discussions among industry practitioners and through more formal working groups.

Recommendations

The overall recommendations of this SAE EDGE™ Research Report include prioritizing the next steps forward. In order to take a resource-leveling approach, these steps have been categorized as immediate, short-term, or medium-term priorities (Figure 7).

The prioritization is designed to balance the resources required to address each proposed step, while pushing high-priority steps to the forefront. "Priority" steps are those that require immediate attention. "Short-term" steps will also require attention in the short term. "Medium-term" priorities should be addressed once priority and near-term steps have been tackled.

Tables 5-7 also summarize the steps under each of the three priority categories. The steps have been renumbered as a series.

Each proposed next step is listed in order of priority, with a brief description of the proposed step as follows:

Priority Steps

1. **Lack of guidance on data sharing and agreements required to support ADS data use**
   This should include advice on the establishment of data governance arrangements that specify what data will and will not be shared and the arrangements for protecting the privacy and security of the data. Ideally this would be developed at a national level to

FIGURE 7. **Aspect prioritization.**

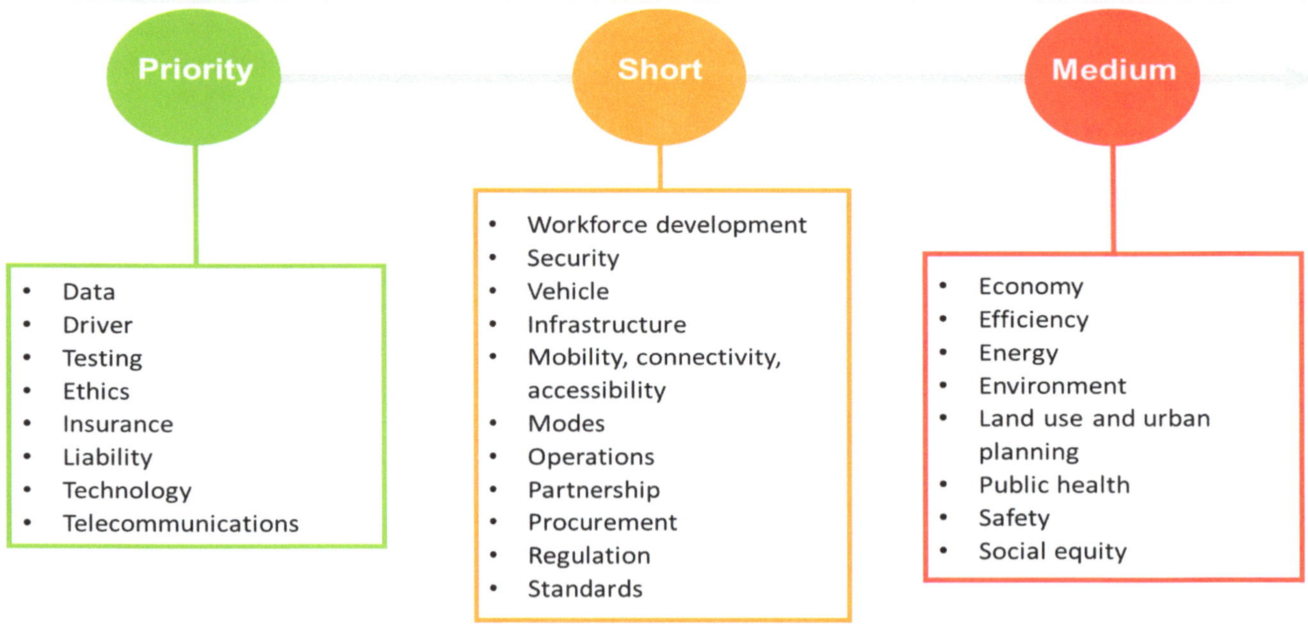

TABLE 5. **Priority steps.**

1	Data	Develop guidance on data sharing and agreements required to support ADS data use
2	Environment	Define the impact of ADS technology and connectivity on the environment
3	Infrastructure	Synchronize infrastructure investments with AV deployment
4	Insurance	Define a detailed assessment of the impact of AVs on insurance rates
5	Liability	Conduct an analysis of liability associated with AVs
6	Operational	Assess the operational impact of AVs on the transportation network
7	Procurement	Review procurement approaches and develop new models that better support effective engagement of the private sector
8	Public health	Understand the opportunities and challenges related to public health and AVs
9	Security	Develop guidance and protocols to ensure the security of data generated from and ingested by ADS-equipped vehicles
10	Standards	Continue work to develop an appropriate set of standards to cover all aspects of ADS design and deployment
11	Telecommunications	Define clear policies that state the need for telecommunications facilities
12	Vehicle	Define policies that preserve public safety while fostering innovation

TABLE 6. **Short-term steps.**

13	Efficiency	Understand the relationship between ADS technology and transportation efficiency
14	Energy	Understand the energy impacts of AVs
15	Ethics	Develop a code of ethics for both the public and private sectors
16	Mobility, connectivity, accessibility	Assess the effects of AVs on mobility, connectivity, and accessibility
17	Partnership	Develop guidance for effective public-private partnership establishment and management
18	Regulation	Develop a full set of regulations and synchronize with existing traffic and transportation
19	Safety	Develop communication tools to explain the safety impacts of AVs to the general public
20	Testing	Ensure the policies governing testing on public roads are consistent and comprehensive
21	Workforce development	Understand the likely impact of AVs on workforce development

TABLE 7. Medium-term steps.

22	Driver	Converge existing policies regarding traditional transportation with ADS policies
23	Economy	Develop a better understanding of the economic impacts of AVs
24	Land use and urban planning	Define the opportunities and challenges associated with land use, urban planning, and AVs
25	Modes	Define the modal impact of AVs
26	Social equity	Ensure the preservation of social equity based on a detailed understanding of the effects of AVs
27	Technology	Ensure that policy and technology capability are properly synchronized

ensure consistency across the various local deployments of AVs. Consideration should also be given to the creation of a national data exchange to enable a bigger picture of data across the nation to be created. This step would support the effective use of data, avoid reinventing the wheel, and accelerate successful ADS application through shared experience.

2. **Need to define the impact of ADS technology on connectivity and the environment**
   In line with energy impacts of AVs, there is also a potential benefit through the reduction of emissions and effects on the environment. A study of the impacts of AVs on the environment along with the improved connectivity provided by such technologies should be conducted. Understanding the ADS impact on connectivity and the environment will provide valuable information to justify investment and communicate to the public the value and benefits of such systems. The information will also show how ADS technology will fit within a wider transportation context.
3. **Need to synchronize infrastructure investments with ADS deployment**
   This should result in a comprehensive framework that illustrates how public sector infrastructure investments will align with and reinforce parallel private sector investments. A study should be conducted that catalogs public sector infrastructure investments and private sector investments in AVs, as well as gaps, potential conflicts, and overlaps. This will enable maximum return on investment from both traditional transportation infrastructure and ADS investments, while ensuring optimal performance from both through alignment and harmonization.
4. **Lack of a detailed assessment of the impact of AVs on insurance rates**
   This should be complemented by a detailed assessment of the likely impact of AVs on insurance premiums. This should cover all modes of transportation and will require a detailed ADS risk assessment. An understanding of the impact on insurance rates is critical to communicating value and benefits to the public. It will also provide significant input into the overall ADS cost-benefit picture.
5. **Need for an analysis of liability associated with AVs**
   While this is likely that the private sector has already conducted an initial analysis of liability, it is important that a publicly available analysis be conducted. There are challenges and opportunities that would benefit from discussion that can be supported by defining and documenting ADS liability issues. The definition of roles and responsibilities and legal liability is a critical aspect of AV operations. This step will provide support for determining operating costs and identification of appropriate measures to manage risk and liability.
6. **Need for an assessment of the operational impact of AVs on the transportation network**
   It is necessary to understand the nature and size of the benefits that can be achieved. Transportation operations consume a considerable amount of resources that govern the cost of moving goods and people. A detailed understanding of the operational impact of ADS technology would support the achievement of economic benefits driven by gains in operational efficiency. These vehicles will not operate in isolation but will form an integrated part of an overall transportation network. This step will ensure that the relationship between conventional and automated transportation is fully understood, enabling the definition of arrangements for optimum operation of both conventional and automated systems.
7. **Need for a summary of procurement approaches and new models that better support effective engagement of the private sector**
   The relative roles and responsibilities of the public and private sectors should be defined, and business models should be created that illustrate who makes the investment and who gains the reward. The balance between public good and private need to develop markets should be encapsulated in the guidance. This will enable public sector agencies to define the appropriate procurement model for the circumstances. This will also support the effective engagement of the private sector in the overall implementation process.

8. **Need to develop an understanding of the opportunities and challenges related to public health and AVs.**
This should take a balanced approach and identify the advantages and disadvantages of AVs. The removal of the driver or operator from a vehicle should enhance public health in terms of reducing interaction and contact. On the other hand, automated vehicles will require an appropriate level of cleaning and sanitizing to protect public health levels. A clear understanding of public health effects will support the definition of optimum operational strategies and business models.
9. **Need for guidance and protocols to ensure the security of data generated from and ingested by AVs**
ADS cybersecurity and physical security should be addressed through the creation of guidance and protocols that preserve the security of data generated from, and ingested by, AVs. Control functions within onboard equipment should be suitably protected to prevent malevolent activity. Such guidance and protocols should also be carefully aligned with current guidance on other aspects of the transportation network and guidance with respect to data management. This will support the definition of plans to ensure the security of ADS data through the definition of data governance and data security arrangements. This will also support the appropriate levels of privacy for system users and other data sources.
10. **Need to develop an appropriate set of standards to cover all aspects of ADS design and deployment**
The standards should cover the design of vehicles and onboard equipment, operating conditions, and requirements for field testing and verification. ADS certification should be considered along the lines of the current driver's license model. The feasibility of issuing a driver's license equivalent to an ADS should be explored. This will support the most effective implementation including economies of scale and the integration of practical lessons learned into future products and services. The definition of standards will also support the establishment of markets and underwrite the future success of the private sector, while minimizing risk to the public sector.
11. **Lack of clear policies that define the need for telecommunications facilities**
These policies should focus on the needs but remain flexible enough to accommodate new and emerging telecommunications technologies. In particular, the public sector role in investing in telecommunications infrastructure and deploying the appropriate equipment should be considered and defined. Telecommunications is a critical component in ADS technology and this step will ensure that related policies adequately address this.
12. **Need for policies that preserve public safety while fostering innovation**
Advice and guidance should be provided for defining policies to preserve public safety while fostering innovation. There is a need for guidance on achieving the balance between preserving public safety and enhancing public safety through the rapid ADS introduction. Such systems have the potential to dramatically reduce accidents and fatalities but need to be implemented to have any effect. This will support an optimum balance between safety, efficiency, and enhanced user experience.

Short-term Steps

13. **Need to understand the relationship between ADS technology and transportation efficiency**
In a similar vein, it is necessary to develop a detailed understanding of how AVs will improve the efficiency of transportation service delivery, particularly in urban areas. While safety is a primary objective, the delivery of a more efficient and effective transportation network is also an important goal to be achieved through AV introduction. It is necessary to understand the mechanisms through which these vehicles will improve efficiency and to at least estimate the size of the efficiency gains that can be achieved. Understanding the impact of AVs on overall transportation efficiency is a critical component in understanding the benefits and costs of such technology. This step will enable more effective cost-benefit analysis and the derivation of communication materials for the public.
14. **Need to understand the energy impacts of AVs**
It is estimated that around 28% of the energy generated in the US is used for transportation [15], providing considerable scope for improvement. ADS energy impacts should also be considered alongside the energy impacts of the entire transportation value chain.
15. **Lack of a code of ethics for both the public and private sector**
Guidance should be provided to both the public and private sector on the ethical aspects of ADS technology. This should address the effect on driver behavior and ethical ADS design and ensure appropriate behavior in both the public and private sector sides with respect to new ADS technology.
16. **Incomplete understanding of the effects of AVs on mobility, connectivity, and accessibility**
Specific areas where ADS technology will impact these important urban transportation parameters should be described and defined. Where possible, the detailed effects of ADS technology interactions should be documented and quantified. This will supply much needed data for a better understanding of effects and a better approach to cost-benefit analysis. It will also support the development of more effective communication materials for the public.
17. **Lack of guidance for effective public-private partnership establishment and management**
This guidance should cover information on how to establish a partnership, how to manage a partnership,

and how to achieve a balance between public and private sector objectives. This will support effective engagement of both the public and private sector resources required to implement ADS technology successfully. It will enable the attainment of public sector and private sector objectives in a balanced win-win approach.

18. **Need for a full set of regulations and synchronize with existing traffic and transportation**
    This should build on existing progress and address the synchronization of ADS regulations with current regulations such as those governing vehicle speed and parking. Such alignment is critical to ensure optimal performance of both.
19. **Need for communication tools to explain the safety impacts of AVs to the general public**
    While an emphasis should be placed on safety, the communication tools should also explain the efficiency and driver experience benefits of AVs. Effective communication to the public will be an important dimension in AV implementation, as it can be used to communicate clear definition of expected benefits, objectives, and quality of life improvements.
20. **Need to ensure that policies governing testing on public roads are consistent and comprehensive**
    There is a need for national, if not global, consistency on this subject. This will enable rapid progress in AV implementation through sharing of practical experiences and lessons learned. Providing a consistent framework at a national level will also support accelerated implementation by making it easier for local agencies to define policies and agree on roles and responsibilities within implementation projects.
21. **Lack of complete understanding of the likely impact of AVs on workforce development**
    This should include an assessment of the skills required for current and future ADS development and the educational needs to support the growth of a new AV industry. While it is helpful for local jurisdictions to have flexibility in developing their own ADS policies, there is also a need for some degree of national alignment of policies. This can be achieved by ensuring that national policies allow sufficient flexibility for local jurisdictions to develop policies that do not conflict, while achieving local transportation objectives. This will support a full and detailed understanding of the likely impacts of ADS technology on workforce development, both positive and negative. This will support the definition of appropriate arrangements for training and professional development and set the scene for education programs to support future needs.

Medium-term Steps

22. **Need to converge existing policies regarding traditional transportation with ADS policies**
    This could be achieved through an analysis of existing policies and a comparison of the contents of the existing policies with those of national policies being developed for ADS. This could also include convergence and alignment of public policy objectives at the various levels of government from national, to state and local.
23. **Lack of a complete understanding of the economic impacts of AVs**
    There is considerable information on the likely size of the ADS market. However, it is necessary to develop information on the likely economic impacts of ADS technology at local and national levels. This should include impacts relating to safety, transportation efficiency, and enhancing the travel experience. It should also provide information on the economic benefits of fostering and supporting an AV industry. This will support a clear and complete understanding of the economic impacts of ADS technology, including direct effects on the user and on the national economy through support for the establishment of new industries.
24. **Need to define the opportunities and challenges associated with land use, urban planning, and AVs**
    Guidance documents and methodologies should be created that explain how to make use of probe data from AVs to improve the effectiveness and efficiency of land use planning and understanding the relationship between land use and transportation demand. This should also include road geometry and other changes such as zoning required in order to support ADS development. Transportation network impacts should also be considered, especially the emergence of transportation mobility hubs that will require adaptation of the current transportation network. This step promotes a clear understanding of the weight of AVs on land use and urban planning.
25. **Need to define the modal impact of AVs**
    This would also help to ensure coherence across transportation modes and avoid any potentially conflicting ADS investments. The emergence of new modes of transportation supported by ADS technology should also be considered as part of this step. ADS technology, combined with other technologies, can provide MaaS opportunities and other innovative approaches based on new technologies and business models. This promotes a balanced approach to ADS introduction based on a complete understanding of the likely impact on all modes of transportation and the emergence of new modes of transportation.
26. **Need to ensure the preservation of social equity based on a detailed understanding of the effects of AVs**
    It is important to have sufficient understanding of the likely impacts to ensure that some sectors of the population are not being disadvantaged through ADS implementation.

27. **Need to ensure that policy and technology capability are properly synchronized**
    This should be used to ensure that policy and technology development are in close alignment. Where necessary policy development should be accelerated to avoid any lag behind technology capability. This supports an alignment between policy and technology capability, bringing them into closer synchronization will ensure that policy takes account of the latest technology capabilities and constraints, while products and services take full account of policy dimensions.

Acronyms

3D - Three-dimensional

ADS - Automated Driving System

AI - Artificial Intelligence

AV - Automated Vehicle

DSRC - Dedicated Short-Range Communications

LIDAR - Light Distance and Ranging

MaaS - Mobility as a Service

mph - Miles per hour

NHTSA - National Highway Traffic Safety Administration

OEM - Original Equipment Manufacturer

Radar - Radio distance and ranging

UK - United Kingdom

US - United States

Acknowledgments

Recognition should go first to all the participants, many of who also provided feedback on the draft version of this publication. Without their input and initiative, this SAE EDGE™ Research Report would not have been possible.

1. Vik Bhide, City of Tampa, Florida
2. Doug Jamison, Central Florida Regional Transportation Authority
3. Jeremy Dilmore, Florida Department of Transportation
4. Peter Thompson, San Diego Association of Governments
5. Barrie Kirk, Canadian Automated Vehicles Centre of Excellence (CAVCOE)
6. TJ Ross, Pace Suburban Bus (Ret.)
7. James Garner, Pace Suburban Bus
8. Peeter Kivestu, Oplytix

The author of this document together with the SAE Team responsible for its creation join in expressing their deepest appreciation to all the individuals mentioned above.

Bob McQueen
Bob McQueen and Associates

References

1. Wikipedia, Locomotive Acts Webpage: https://en.wikipedia.org/wiki/Locomotive_Acts, accessed March 20, 2020.
2. Merriam-Webster Dictionary: https://www.merriam-webster.com/dictionary/policy, accessed March 20, 2020.
3. Society of Automotive Engineers, SAE Mobilus, Unsettled Topics Concerning the Field Testing of Automated Driving Systems, EPR 2019009, https://www.sae.org/publications/technical-papers/content/epr2019009/, accessed March 20, 2020.
4. Industry Week, The Automotive Industry: Economic Impact and Location Issues Webpage: https://www.industryweek.com/the-economy/article/21958422/the-automotive-industry-economic-impact-and-location-issues, accessed March 20, 2020.
5. Center for Automotive Research, Contribution of the Automotive Industry to the Economies of All 50 States and the United States, Apr. 2010: https://www.cargroup.org/publication/contribution-of-the-automotive-industry-to-the-economies-of-all-fifty-state-and-the-united-states/, accessed March 20, 2020.
6. CITYLAB, Traffic's Mind-Boggling Economic Toll, Feb. 7, 2018: https://www.citylab.com/transportation/2018/02/traffics-mind-boggling-economic-toll/552488/, accessed March 20, 2020.
7. Driverless Cars and Accessibility, Designing the Future of Transportation for People with Disabilities, ITS America, Apr. 2019: https://static1.squarespace.com/static/596fb16003596e0fa70a232f/t/5c9bab319b747a61663ac9bc/1553705778370/ITSAmerica_Driverless+Cars+Accessiblity+Mobility_April2019.pdf, accessed March 20, 2020.
8. Optimezely, Optipedia, Optimization Glossary, A/B Testing Page: https://www.optimizely.com/optimization-glossary/ab-testing/, accessed March 20, 2020.
9. National Highway and Traffic Safety Administration, Automated Vehicles Webpage: https://one.nhtsa.gov/Research/Crash-Avoidance/Automated-Vehicles, accessed March 20, 2020.
10. National Highway and Traffic Safety Administration, About NHTSA Webpage: https://www.nhtsa.gov/about-nhtsa, accessed March 20, 2020.

11. Federal Reserve Bank of St. Louis, What Is Workforce Development, Apr. 1, 2010: https://www.nhtsa.gov/about-nhtsa, accessed March 20, 2020.
12. University of Michigan Transportation Research Institute, Travelers Vehicles: Fewer Cars, More Miles, Feb. 11, 2015: http://www.umtri.umich.edu/what-were-doing/news/driverless-vehicles-fewer-cars-more-miles, accessed March 20, 2020.
13. The Washington Post, Five Myths about Autonomous Vehicles: https://www.washingtonpost.com/outlook/five-myths/five-myths-about-autonomous-vehicles/2019/08/15/245c39bc-bec6-11e9-b873-63ace636af08_story.html, accessed March 20, 2020.
14. Brookings Institute, Gauging Investment in Self Driving Cars, Oct. 16, 2017: https://www.brookings.edu/research/gauging-investment-in-self-driving-cars/, accessed March 20, 2020.
15. U.S. Energy Information Administration, Use of Energy Explained: https://www.eia.gov/energyexplained/use-of-energy/transportation.php, accessed March 20, 2020.

Sources of Further Information

This SAE EDGE™ Research Report draws on an extensive set of information sources related to the ADS policy. In the course of developing the report, it became obvious that there is a significant body of knowledge that already exists. The purpose of this report is not to duplicate or replicate the body of knowledge, rather synthesize current initiatives and develop an understanding of unsettled issues and further work that would be required. In order to serve as a single repository for information regarding policy aspects, the following list of additional information sources is provided.

1. Ensuring American leadership in Automated Vehicle Technologies, Automated Vehicles 4.0, National Science And Technology Council and the US Department of Transportation, January 2020 http Auch Martins://www.transportation.gov/policy-initiatives/automated-vehicles/av-40
2. Law Commission United Kingdom, webpage on Automated Vehicles: https://www.lawcom.gov.uk/project/automated-vehicles/
3. Center for Connected and Autonomous Vehicles, UK webpage, https://www.gov.uk/government/organisations/centre-for-connected-and-autonomous-vehicles
4. National Highway Traffic Safety Administration, Automated Vehicles for Safety webpage: https://www.nhtsa.gov/technology-innovation/automated-vehicles-safety
5. Automated Vehicles, a Joint Preliminary Consultation Paper, Law Commission United Kingdom, Scottish Law Commission, November 8 2018: https://s3-eu-west-2.amazonaws.com/lawcom-prod-storage-11jsxou24uy7q/uploads/2018/11/6.5066_LC_AV-Consultation-Paper-5-November_061118_WEB-1.pdf
6. Automated Vehicles: Consultation Paper to on Passenger Services and Public Transport, a Joint Consultation Paper, Law Commission UK and Scottish Law Commission, October 16, 2019: https://s3-eu-west-2.amazonaws.com/lawcom-prod-storage-11jsxou24uy7q/uploads/2019/10/Automated-Vehicles-Consultation-Paper-final.pdf
7. United States Department of Transportation, Automated Vehicle Activities webpage, https://www.transportation.gov/AV
8. Society of Automotive Engineers, Automated and Unmanned Vehicles webpage, https://www.sae.org/automated-unmanned-vehicles/
9. US Chamber of Commerce, Chamber Technology Engagement Center, Automated Vehicle Policy Principles: https://www.uschamber.com/issue-brief/ctecs-automated-vehicle-policy-principles
10. US Chamber of Commerce, Chamber Technology Engagement Center, Data for Good: Promoting Safety, Health, and Inclusion: https://americaninnovators.com/research/data-for-good-promoting-safety-health-and-inclusion/
11. Intelligent Transportation Society of America, Automated Vehicles webpage: https://www.itsa.org/policy-autonomous-vehicles
12. European Commission, Automated and Connected Mobility in Europe, Policy webpage: https://ec.europa.eu/digital-single-market/en/connected-and-automated-mobility-europe
13. Automated Driving, Roadmap for the Deployment of Automated Driving in the European Union, European Automobile Manufacturers Association: https://www.acea.be/uploads/publications/ACEA_Automated_Driving_Roadmap.pdf
14. Autonomous Vehicle Implementation Predictions, Implications for Transport Planning, March 24, 2020, Victoria Transport Policy Institute: https://www.vtpi.org/avip.pdf
15. Stick Shift: Autonomous Vehicles, Driving Jobs, and the Future of Work, Center for Global Policy Solutions, March 2017: http://globalpolicysolutions.org/report/stick-shift-autonomous-vehicles-driving-jobs-and-the-future-of-work/
16. The Nature Of Transport Policy, Dr. Brian Slack, Dr. Theo Notteboom and Dr. Jean-Paul Rodrigue, Routledge New York 2020, https://transportgeography.org/?page_id=6279
17. Principles For Autonomous Vehicle Policy, American Planning Association, https://www.planning.org/policy/principles/av/

18. Bipartisan Center, What's Happening with Autonomous Vehicle Policy?, July 16, 2018:: https://bipartisanpolicy.org/blog/whats-happening-with-autonomous-vehicle-policy/
19. Autonomous Vehicles Policy Knowledge Toolkit, Harvard University, https://www.hks.harvard.edu/centers/taubman/programs-research/autonomous-vehicles-policy-initiative/autonomous-vehicles-policy-knowledge-toolkit
20. Autonomous Vehicle Technology-A Guide For Policymakers, RAND Corp., https://www.rand.org/pubs/research_reports/RR443-2.html
21. The Hill, Stakeholders Must Engage Policymakers on Driverless Cars: https://thehill.com/blogs/congress-blog/technology/306670-stakeholders-must-engage-policymakers-on-driverless-cars
22. The Robot Report,Stakeholders Likely to Be Disrupted by the Changeover to Self Driving Vehicles: https://www.therobotreport.com/stakeholders-likely-to-be-disrupted-by-the-changeover-to-self-driving-vehicles/
23. Deploying Autonomous Vehicles, Commercial Considerations and Urban Mobility Scenarios, Ernst & Young: https://www.ey.com/Publication/vwLUAssets/EY-Deploying-autonomous-vehicles-30May14/%24File/EY-Deploying-autonomous-vehicles-30May14.pdf

Contact Information

EDGEresearch@sae.org.

www.ingramcontent.com/pod-product-compliance
Lightning Source LLC
LaVergne TN
LVHW070937210926
848482LV00003B/263

* 9 7 8 1 4 6 8 6 0 2 0 4 3 *